(2019.)

To.

OUR DARLIN[...]

LOTS OF L[...]

AUNTIE

NANNY & JO JO
  Xx   xx

HAPPY READING MY SWEETIE
            "OUR"
                    Xx

12
SMALL
ACTS TO
SAVE
OUR
WORLD

WWF

12
SMALL
ACTS TO
SAVE
OUR
WORLD

CENTURY

1 3 5 7 9 10 8 6 4 2
Century
20 Vauxhall Bridge Road
London SW1V 2SA

Century is part of the Penguin Random House group of companies
whose addresses can be found at global.penguinrandomhouse.com

Penguin
Random House
UK

Text by Emily Beament
Illustrations © Adam Doughty 2018
adamdoughty.com

©1986 panda symbol and ® "WWF" Registered Trademark of WWF.
WWF-UK registered charity (1081247) and in Scotland (SC039593).
A company limited by guarantee (4016725)

First published in the United Kingdom by Century in 2018

www.penguin.co.uk

A CIP catalogue record for this book is available from the British Library

ISBN 9781780899282

Typeset in 11/15 pt Avenir by Jouve (UK), Milton Keynes
Printed and bound in Great Britain by Clays Ltd, Elcograf S.p.A.

Penguin Random House is committed to a sustainable future for
our business, our readers and our planet. This book is made from
Forest Stewardship Council® certified paper.

MIX
Paper from
responsible sources
FSC® C018179

Contents

Foreword by Ben Fogle, WWF ambassador and UN Patron of the Wilderness

Despite what Hollywood might have you believe, you don't have to be the Black Panther, Thor or Wonder Woman to save our planet.

Who among us doesn't want to clean up our oceans, protect the polar regions and lower our plastic footprint? Who doesn't want to avoid the worst impacts of climate change, eat sustainable food and stop the destruction of vital habitats for the world's amazing animals?

And who wouldn't want to fight and save the day like the Caped Crusader?

I'm fortunate enough to have travelled to every corner of the globe and I never cease to be amazed, astounded and thrilled by the sheer beauty and brilliance of our shared home. But I've also become increasingly enraged and distressed by the damage we're doing and the speed at which we're doing it.

We're the first generation to fully understand the havoc humans are causing. We're also the last generation who can turn things around and make things right – for the health, wealth and security of our children, grandchildren and generations to come.

Many of us find issues such as climate change and deforestation daunting, and solutions to overcome them even more confusing. What can I really do in my daily life that will make a difference? What's the point in me recycling or not using plastic straws if big businesses refuse to change? Where do I begin?

Well, the book you find in your hands is for all of us – the everyday heroes who have to do a weekly shop, make packed lunches for our kids and battle the demands of modern living. It's written in collaboration with the great people at WWF, and it's full of small, simple, achievable acts that together can have a great impact on saving our planet. You can read it wearing a mask and cape if you like.

It comes as little surprise that with the fast pace of twenty-first-century life we have become used to some shortcuts, and we've picked up bad habits. The everyday plastic food packaging and water bottles are designed to keep our produce fresh for longer. The competition to supply us with the cheapest clothes or energy can keep our bills down. But these conveniences come at a cost: they're doing irreversible damage to our environment. It's a cost we can no longer continue to ignore.

I am constantly struck by the stunning natural beauty of our world. I have trekked across the Arctic, rowed the Atlantic, and recently I fulfilled a childhood dream when I climbed to the roof of the world and stared in awe from the peak of Everest. It's at times like these that I really grasp how fragile our ecosystems are, and appreciate how we must do everything to protect our world, and our place in it. As a WWF ambassador and UN Patron of the Wilderness, I try to share my experiences, shedding light on success stories as well as the areas for urgent change. I want my children and yours to be free to explore and discover the world like I have.

Before you begin this book, I want to urge you that it's never too late. My recent expedition to Everest was an inspiring example of this. Before the trip, people kept asking, 'Isn't it covered in rubbish these days?' But the Everest I discovered was nearly back to its stunning best. Despite getting 100,000 visitors a year, it was one of the cleanest, tidiest wilderness trails I have encountered.

This is thanks to a number of things, including a major clean-up operation by the Nepalese government, strict regulations for all climbers, and a litter ban that has recently been introduced. Small acts that together have had a tremendous impact.

Now really is the time to act. You don't even have to be a superhero – everyone can make a difference. By reading about the twelve acts presented in this book, and just adjusting your lifestyle to fit one, two, three – or, I really hope, all twelve of them – you will be making a difference. A difference that will be felt by you, your community and your world. By buying this book you've already taken an excellent step towards doing your bit.

So let's all join the fight to save our planet. It's our time to turn things around and actively restore nature rather than selfishly taking from it. Let's be the custodians our world deserves.

And who knows, maybe one day Hollywood will be making films about us and how we stood up and saved the world.

Introduction

Do you ever wonder about how much the world has changed during your lifetime, from the technology that has altered how you live to the environmental changes taking place in our oceans and atmosphere? What about your place on this planet, and how much your actions affect it? Do you often feel like a helpless bystander when it comes to global challenges?

We only need to switch on the TV to be reminded of the terrifying changes we are witnessing: glaciers are melting as temperatures rise; there are more floods and droughts; seas are swimming with plastic; the forests are disappearing; and the rate at which species are going extinct is climbing like a heart monitor for a patient under intense stress. All too often the TV news feels like the opening scenes of a disaster movie.

It is easy to feel like these changes are irreversible, that you don't have any control over what is going on in the world and can't make a difference. We understand. But trust us; you have a huge part to play. The environment is not something out there, too big a matter for an individual to alter. Instead, we are all part of it, and every one of our choices – no matter how big or small they seem – has an inextricable impact on the world around us.

With this book, we'll put you back in control, and help you take some simple steps in your own life to steer the world in the right direction and secure a better future for the people and all the other species that live on this planet.

We're not just presenting you with a list of dos and don'ts to tick off. This book is brimming with wonderful ideas from experts and examples of how people around the world are transforming their own lives and the lives of others. It aims to show how you can take meaningful action in your daily life and activities rather than just standing by, from the food you purchase to how you get to the office. It also explains why it is important that we all take action, and how your first act can be a starting point for a whole host of changes you can make in your life.

There's an added bonus, too – as well as being good for our oceans, our atmosphere, our rivers and our land, these changes are pretty likely to make you feel healthier, more comfortable and better off at the same time. They might even make you feel a little happier, and should certainly make you think a little differently about our world and your place in it.

The changes here begin at home, at the office, and in the places that our daily lives take us. But they don't end there. Individual actions can send a message to our leaders in governments and to corporate boardrooms that there's a larger movement – a groundswell of people who want those in power to enact change on a planetary scale. Your actions, whether they are as simple as carrying a reusable water bottle or using both sides of the paper when printing at work, might encourage others to do the same, and *can* inspire real change.

Just think back for a moment, ten or twenty years back. What was there in the way of recycling facilities in your area? Maybe you made an occasional trip to the bottle bank, listening to the glass smash and shatter as you posted the bottles through the hole one at a time, but not much more than that. And now, though it can still be hugely improved, opportunities to recycle are nearly everywhere we look; millions upon millions of cans, bottles, cardboard containers and all the rest are recycled every day, saving resources, saving energy, saving them ending up in landfill. A positive change in direction can take place in just a few years, driven by people, like you, who want to see that change.

Or think of how some species of wildlife pushed to the brink by human activities have been helped to reverse their fortunes. Otters vanished from much of England when their rivers were poisoned by farming chemicals; now, slowly but surely, after the chemicals were controlled, some rivers were cleaned up and the otters given new protection status, they are increasing in number once again. Condors driven almost to extinction by shooting, poisoning and the destruction of their habitat are back in the skies of California after a captive breeding and reintroduction programme.

Now try to imagine the world in twenty or thirty years' time, restored by the changes we could make: cleaner air, greener cities, a countryside alive with birds, bees and butterflies, healthy rivers and forests around the world still acting as the planet's lungs, teeming with wildlife, a stable climate for your children and their children. It's not as far-fetched as you might think, but it helps to know where to start along the path to this better, more sustainable world.

So what are you waiting for? Come on, follow us, and we can change the world together, one small act at a time.

Don't simply standby: Unplug your appliances and kill the vampire power

They've got little glowing eyes, red, green or blue, that follow you in the darkness as you leave the room at night, heading for bed, with a yawn and a quick glance around to make sure everything is as it should be before you go to sleep. In the morning when you come back in, still sleepy, and rubbing your eyes, they gleam at you in the half-light from their place crouching in corners or on side tables, before you open the blinds or curtains.

They've been glowing away in all the dark rooms of your house while you sleep. And when you go off to work, they're going to carry on glowing, greedily pulling electricity out of the mains through their plugs while they wait for you to come back and use them, or quietly telling the time to empty rooms as the day passes.

When you leave appliances such as TVs, microwaves, laptops and phone chargers plugged in and on standby, it's like there's a host of little vampires around your home sucking away, using electricity and adding to your bills without you benefiting at all.

Over the space of a year, this wasted electricity mounts up. Even if it's an energy-efficient TV, keeping it on standby all year uses as much power as leaving a light bulb on for a day and half. Keeping your games console on standby throughout the year is like leaving a light permanently on for almost the whole of the year, according to the Energy Saving Trust.

So when you go to bed at night, or before you leave for work in the morning, see off the vampires. Check round the house and switch the appliances off at the wall or unplug them. Make sure your phone charger isn't left plugged in. Even if you just did this for one or two appliances, where it's convenient, it would make a difference, as each one is using electricity, wasting money and the natural resources used to create the power.

When it comes to saving energy, every little helps. And saving energy is a big part of what we need to do to solve the biggest challenge of our time: climate change.

It can be hard to think about how our actions affect the climate. What can one person do in the face of rising temperatures caused by putting increasing amounts of greenhouse gases into the atmosphere, and the threat of heatwaves, floods, rising sea levels, more intense storms and damage to crops and wildlife? It's no wonder that some people choose not to think about it, or even that it's not happening.

But we are on the cusp of a revolution. Countries, recognising the need for action, struck a deal in France in 2015 to prevent dangerous climate change. Under this deal, the Paris Agreement, countries will take steps to cut their greenhouse gas emissions, which come from things like burning fossil fuels in power plants and car engines and cutting down forests. This aims to keep global temperature rises to 'well below' 2 degrees Celsius higher than they were in the nineteenth century, before the world really started to industrialise. It also promises to try to limit the temperature rise even further to 1.5 degrees Celsius. That's because many countries, such as low-lying islands in the Pacific, fear that once we get past 1.5 degrees of warming, things are looking pretty bleak for them.

But here's the thing. To halt temperature rises at 2 or 1.5 degrees Celsius or to stabilise the climate at any temperature, we need to seriously cut back or stop putting more greenhouse gases such as carbon dioxide into the atmosphere. Emissions from every source, from energy to transport to cutting down forests, have to be brought to zero (or as near to zero as possible) overall, while any greenhouse gases that can't be avoided need to be cancelled out by doing things like planting more forests. To avoid dangerous climate change, we have to get to a net zero-emissions world before mid-century – and the earlier the better.

Burning fossil fuels to power and heat homes, drive industry, provide the food we eat and run our transport systems makes up the majority of the billions of tonnes of carbon we put into the atmosphere each year, and generating electricity is a big part of that.

But that's where the good news begins: we're already beginning to see a shift towards cleaner power generation, with tumbling prices for renewable technologies such as solar panels. And wind turbines, and lots of countries around the world are rolling out these technologies at an extraordinary rate-including huge economies like China, but also the UK. In fact, on Friday 21 April 2017, the UK – birthplace of the Industrial Revolution – had its first full day without any electricity generated from coal since it started using the fuel to generate electricity back in the 1880s. We've already had a streak of three consecutive such days in 2018. Meanwhile, the UK's use of coal for power generation has dropped from 40 per cent to less than 7 per cent in the last five years.

Between 2014 and 2016, the world saw either little growth in global emissions or no growth at all for three years in a row, suggesting that greenhouse gas output could finally be plateauing. And although emissions rose again in 2017, evidence suggests we can turn things around again.

A big piece of the puzzle is using less energy. The International Energy Agency describes energy saving as the one energy resource every country has in abundance, and the first fuel of economic development. It is by far the cheapest and quickest way of tackling the costs of energy, making sure supplies are secure and cutting carbon.

And everyone can play their part in saving energy. The simplest way is unplugging appliances or switching them off at the wall. Once you've got into that habit, there are lots of other things you can do to save energy around the home. If you boil a kettle, only fill it with as much water as you need. If it's just for one cup of tea or coffee, keep the amount of water to the minimum, and use the indicators on the side of the kettle as a guide. If you're making tea, this doesn't just save energy and money, it also makes for a nicer cuppa, as experts say you shouldn't make tea with water that's already been boiled, because it spoils the flavour.

Light bulbs have got much more efficient, too, and with the advent of the latest generation of LEDs (or light-emitting diodes), you can have a choice of brightness, colour and even the capability to dim the lights, which was once only possible with old-style light bulbs. Prices of LEDs are falling fast, they use much less electricity and so are cheaper to run, and once they're in, you don't need to replace them for years. And if you still really love the look of a traditional Edison light bulb with the glowing filaments, guess what? You can get LEDs of those, too.

Once you think you've done all the things you can to save electricity around the home, think again about where your electricity comes from in the first place. By the second half of the century (and decades sooner in some countries), all that power is going to have to come from low-carbon sources such as wind farms. In just a couple of minutes, you could get ahead of the game, and switch to a 100 per cent green energy tariff to support companies that are already driving this shift.

The onus is all on the providers to get it done, making switching easy, and it will save you money. You could even get your own solar panels for power, if you want to be your own provider.

It's not just about electricity, of course, because heating is a major source of energy usage. If you've got a gas boiler, or you're off-grid and you're using an oil burner, or if you have an open fire or stove where you burn coal, you're heating your home with a fossil fuel that produces carbon dioxide. In some ways the use of fossil fuels in heating is a harder challenge to tackle than cleaning up electricity, because instead of big dirty power stations which can be swapped for wind or solar farms, there's a little dirty power station in everyone's home. Things like gas boilers have got more efficient, which in the UK has led to the fall in gas use by households, but they are still producing emissions.

There are lots of ways of tackling this. Gas for boilers could be replaced by injecting renewable 'green gas', made from composting food or plant material, into the grid. Or it could be swapped with hydrogen that has been made from natural gas, in a process that gives off carbon but then could capture this carbon and store it permanently underground. The hydrogen could then be piped into people's homes, where using it for heating only produces water as a by-product.

Then there are air-source heat pumps that are powered by electricity and work like a fridge in reverse, extracting heat from the outside air to generate heat for the home. Similar technology, which can also use temperature differences between the air and water or the ground, can be used on a bigger scale in district heating systems where dozens or even hundreds of buildings can be connected up via pipes to a single energy source which carries heat into homes, or cools them down. In Paris, the River Seine is being used to cool buildings across the city, including the Louvre museum and the National Assembly.

Whichever way you look at it, using less energy for heating is a really good way to start, and that's something that householders can get going on long before anyone starts digging up your street.

You can start with really simple things, like drawing the curtains at night, to keep out the cold and make your rooms feel snug. If you have any spare rooms, turn down the radiator in there: you won't notice and it will save money and energy on heating. If your house is already nice and warm, you can think about turning down the thermostat by 1°C. You probably won't even feel the difference.

Other options include getting double glazing, which helpfully reduces noise if you live on a particularly busy street. You should also make sure your loft and any under-floor cavities are well insulated, which won't break the bank should they need improving, and put insulating strips around your doors if they're draughty. Hold up your hand to the side of the door on a cold, breezy day and if you can feel air moving, that's a pretty sure sign you could do something to improve how well the doorway in insulated.

The great thing about making sure your house is insulated is that it's not just good for keeping the home warm when it's cold, it'll also help keep it cool when the weather is hot, making your life more comfortable, keeping you healthier and saving you money on your energy bills at the same time as helping the world.

Climate change - what's it all about?

It was back in 1824 that French scientist Joseph Fourier worked out that the atmosphere acts like a blanket, keeping the Earth warmer than it would be otherwise, which is commonly known as the 'greenhouse effect'. In 1861, in his lab in London, physicist John Tyndall measured how certain gases can trap heat, including carbon dioxide, methane and nitrous oxide.

The effect these gases naturally have on the planet is a good one – without them, the Earth would be about 30 degrees Celsius colder than it is, and wouldn't be able to sustain life as we know it.

But since the dawn of the Industrial Revolution, humans have been putting more and more carbon dioxide into the atmosphere, by burning coal, gas and oil for industry, transport, heating and electricity. We've also been cutting down forests at an increasing rate and turning ever more land over to agriculture, creating more greenhouse gas emissions.

And once the extra greenhouse gases are in the atmosphere, they do what they've always done best: trap heat.

The potential impact of extra greenhouse gases on global temperatures has been recognised for more than a century, and in 1938 steam engineer and amateur meteorologist ➡

Guy Callendar published the first evidence that the world had recently warmed, largely due to extra carbon dioxide.

In 1958 Charles David Keeling started recording levels of carbon dioxide, which makes up around 85 per cent of total greenhouse gas emissions, at a remote monitoring station in Hawaii. These measurements have been taken ever since, and the resulting 'Keeling Curve' is a clear record of the unrelenting rise of carbon dioxide in the atmosphere.

As carbon dioxide levels have continued to rise, so too have temperatures. Scientific bodies around the world measure global temperatures using data from across the surface of the planet. They've found that the global average temperature has risen by around 1°C since the nineteenth century, and while some years can be a bit colder or warmer, the long-term trend is of a warming world.

All the research that scientists are doing into the impacts of these rising temperatures is assessed by a global organisation appointed by governments – the Intergovernmental Panel on Climate Change – which has concluded that the warming of the planet is unequivocal, the main cause is human activity, and without rapid action the impacts will increase.

Much of the extra heat ends up in the ocean, making it expand and pushing up sea levels, which are also rising because of melting ice sheets and glaciers. We're already seeing sea levels rise dramatically – 20cm between 1901 and 2015.

In some areas, rising temperatures will lead to droughts and heatwaves, while a warmer atmosphere can also hold more moisture, leading to a likely increase in extreme storms and rainfall, causing problems such as flooding.

Harvests of crops such as wheat and rice are likely to suffer, and many species of wildlife already struggle to adapt and will do more in future as traditional seasons are disrupted or the places they live change in ways that make them unsuitable for the animals and plants that live there.

And there is the risk of hitting 'tipping points' – sudden and extreme events such as the melting of the permafrost in the Arctic, which could release masses of methane into the atmosphere – changing our world for ever.

What's your carbon footprint?

While it sounds like a sooty mark on the ground, your carbon footprint is actually a way of measuring how much of an impression you're leaving on the planet. It looks at all the greenhouse gas emissions you generate, and lets you compare that to national and international ➡

averages – and understand how much we all need to muck in if we're to do our bit to tackle climate change.

The footprint covers the emissions associated with the energy you use personally for electricity, for heating your home and for travel, as well as the emissions that are required to produce your food and the other stuff you buy, whether it's made in the UK or somewhere else in the world.

Though it's called the carbon footprint, it includes the various different greenhouse gases, including methane and nitrous oxide, which have different levels of impact on the climate. For example, although less methane goes into the atmosphere each year than carbon dioxide, it's more potent when it comes to global warming, so 1 tonne of methane emissions works like 25 tonnes of carbon dioxide.

A footprint calculator takes that into account, changing all the greenhouse gases into their equivalent amounts in carbon dioxide to give you a single figure (so when you see something described in terms of a 'tonne of carbon dioxide equivalent', or CO_2e, that's what has been done).

Calculations also factor in where you live, because different governments have different policies which affect how you measure up. For instance, if you drive an electric car in Norway, where most electricity comes from renewable ➡

hydropower, that's going to have a smaller carbon footprint than driving an electric car in Britain, where a significant chunk of power still comes from fossil fuels (unless you're already using a clean provider or are on a renewables tariff, of course).

In the UK the average citizen has a carbon footprint that is more than double the global average, while for people in the US it is almost five times higher than the average across the world. Many people in the world's poorest countries – for example those who don't have access to grid electricity, don't drive, and eat food they grow themselves – have negligible carbon footprints.

If everyone had a 'fair share' of the carbon emissions humans will still be able to put into the atmosphere in 2050 and meet the targets to curb temperature rises, it would equate to 1.05 tonnes a year for each of us, far lower than current levels for Western countries. Some countries are taking it even further; Sweden has already committed to net zero.

To check out the WWF carbon footprint calculator, you can head online to: wwf.org.uk/footprint

The future's bright . . . and windy

Before the Industrial Revolution, renewable sources of energy were common, whether it was watermills and windmills, wood for heating and cooking, or animal fats for lights. Then came coal, oil and gas, and much of the world switched to these fossil fuels for their efficiency and ease. But now, with the combination of our understanding about what damage fossil fuels are doing, and leaps forward in technology, we may be at a tipping point where wind, water and the sun replace dirty fuels as the power sources of the future.

In 2016, global renewable electricity generation grew by an estimated 6 per cent and represented around 24 per cent of the world's power output, according to the International Energy Agency. Around 70 per cent of that came from hydropower, mostly large-scale dams which trap water and put it through turbines to generate electricity. Bioenergy accounted for 9 per cent, while wind and solar panels checked in at 16 per cent and 5 per cent respectively.

Across the world almost two-thirds of new electricity resources being installed are renewable, and wind and solar panels which generate electricity directly from the sun are now playing a big part in that.

Mass roll-out and government support is causing prices to fall dramatically and that in turn is helping boost the installation of more renewable tech.

Denmark, a pioneer in modern wind turbines, got almost 44 per cent of its power from wind in 2017. In Europe, many countries are looking offshore for their wind farms, where the newest turbines stand taller than London's Gherkin skyscraper and double the height of the Statue of Liberty. They can power a home for twenty-nine hours with a single sweep of their blades.

Support for the technology has seen costs tumble. Recent auctions to secure contracts for supplying electricity from new offshore wind farms in the UK saw prices more than halve in two and a half years, putting offshore wind on a par with gas plants and making it cheaper than new nuclear power. The UK is now the single biggest market in the world for energy from offshore wind.

Across the US there are more than 54,000 wind turbines, enough to power 27 million homes, according to the American Wind Energy Association, with Texas leading the way among the states, and new wind farms are being built.

But dwarfing everybody is China, which, despite its large amounts of coal, is installing thousands of wind turbines a year, along with huge numbers of solar panels.

China is not alone in rolling out large numbers of solar panels, with vast solar parks being built in the US and India among others – though

China is perhaps the only country with a solar farm in the shape of a panda. In India recent auctions produced some of the world's lowest prices for the technology.

Solar isn't just useful at scale, it's also a way of bringing power to poor and rural communities who are currently not on national electricity grids. The International Energy Agency estimates that by 2022 an extra 70 million people in Asia and sub-Saharan Africa will have electricity thanks to solar home systems.

Small-scale solar is a 'leapfrog' technology – like mobile phones, which have given people communications access without needing to install landline cables. Solar lamps are a good example of this: they work without having to plug people into a major grid infrastructure, and can replace expensive and unhealthy kerosene used for lighting in poorer rural communities, allowing children to do their homework in the evenings, and midwives to see to deliver babies, and even making it safer for women and girls to visit communal toilets at night.

One of the concerns about a greater reliance on renewables to power nations is the fear of 'intermittency' – the sun does not always shine, and the wind does not always blow. But alongside efforts to ensure a diverse supply of clean technology, improve energy efficiency and balance supplies with demand better, there is also the growing role of battery technology, ➡

which can store power when there is excess supply and release it to meet peak demand. Here again, prices are tumbling.

The founder of electric vehicle company Tesla, Elon Musk, has delivered on his promise to install a huge 100 megawatt battery storage unit in South Australia to help tackle problems with power supplies. The installation can power up to 30,000 homes for an hour, and helps balance the grid when demand is high.

It is the world's largest battery unit, but with many more being planned and rolled out, it won't be the biggest for long.

2

Go blue to stay green: Save water each day brushing and washing

It's such a part of the morning routine, and the things you do before bed, you probably don't even think about it. Turn on the tap, wet the toothbrush, put on some toothpaste and get scrubbing.

During those two minutes of cleaning you might be thinking about the day ahead or the day that has just gone. Perhaps you're making a mental to-do list, or checking the weather to see if you need a coat. Maybe you've not quite woken up or you're already half asleep.

But if you've left the tap on while you're teeth-cleaning and daydreaming, you'll have let pints and pints of water go pouring, unused, down the drain. To see how quickly water flows through your taps, you can get a measuring jug from the kitchen and watch it fill up – it'll be overflowing in seconds.

A running tap can waste 6 litres or 10.5 pints of water in a minute according to Waterwise, an organisation that focuses on reducing water consumption in the UK. So if you're brushing your teeth for the recommended two minutes, twice a day, every day, and you're not switching off the water while you scrub, that's 24 litres washing down the drain a day, 168 litres a week, and more than 8,700 litres of water wasted in a year. You could soak in the tub a hundred times a year, or twice a week in nice bubble baths, and use less water than that.

If the hot tap is on too, so you're not cleaning your teeth in icy-cold water on a chilly day, all the energy that you've wasted heating it up, and the cash you have to splash for that, is circling down the drain. Even if you're just using cold water, it will have taken energy to treat the water so it's fit for human consumption, and to pump it to your house in the first place. Taking a second to turn off the tap with one hand while you wield your toothbrush with the other can save energy, money and water.

And when you've finished, make sure the tap is turned right off, because every drip counts.

Water is very precious, wherever you are in the world.

We think of our world as a 'blue planet' – and it is, with oceans covering the majority of the globe, rain-filled clouds swirling overhead, and great cities sited on the

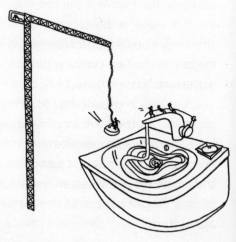

edge of rivers. Over the course of history we have built some of our great civilisations along the course of rivers and lakes, from the ancient Egyptians on the Nile to Lake Titicaca, home of the Incas. We have been drawn to settle by these waters for the natural riches they can give us, green fields and fertile land, water to support homes and industry, and waterways to transport goods and people.

But only a very small amount, less than 3 per cent, of the water on the planet is fresh water, rather than salty, and less than 1 per cent of that is available for people to use, because the rest is locked up in ice and snow, underground or even in permanent frost in the ground. Just a tiny fraction of the world's fresh water is in the millions of lakes and rivers scattered across the continents, and a quarter of river basins run dry before they reach the sea.

Many people rely on groundwater for their water sources. And while a person needs around 20 to 50 litres of water a day to meet their basic needs for drinking, cooking and washing, demand has grown hugely in the last century, particularly in cities, where water use has increased five-fold since 1950.

Population growth has put more pressure on water supplies. So too have improved living standards, which have made water more easily available in people's homes. And it's not just people washing and drinking – the crops we grow use an awful lot of water too. In fact, 70 per cent of all the water we use goes to agriculture.

As a result, around a third of the world's population live with a severe lack of water for at least one month a year, around half of them in India and China. More than one in six don't have access to clean water, and unsafe water is a bigger killer than all forms of violence, including war.

And when we're done with it, 80 per cent of the world's waste water, which includes sewage and the used water from industry, just gets dumped untreated into oceans, lakes and rivers.

With climate change, water is likely to become scarcer in hotter, drier areas. Shifting rainfall patterns could see more dramatic swings between floods and droughts. Melting glaciers which feed some of the world's great rivers could make water supplies much more unpredictable downstream.

By 2030 it is estimated that almost half the entire population of the world will be living in areas where there just isn't enough water.

Rivers, lakes and wetlands are hugely important to wildlife too. They create habitats that are home to species ranging from river dolphins and manatees to thousands of different types of fish, amphibians, birds and insects. These habitats support around a tenth of the Earth's species in all, and provide food, protection from flooding, carbon storage and other benefits to people as well as water supplies.

But water supply isn't only a problem in places that are hot and dry. Hundreds of cities around the world, from London to Tokyo and Miami to Moscow, face challenges over where they get their water.

England's Environment Agency has warned that already overstretched water resources, combined with climate change and a growing population, could see significant supply shortages by the 2050s, particularly in the south-east of the country, hurting people and wildlife alike.

So even if you live in a place where you think there is plenty of water, saving water is something that will help wildlife and wild places, as well as cut the energy needed to treat water to safe standards. After all, it doesn't just come from a tap.

Your teeth should be clean by now, but before you've left the bathroom, there are lots of other things you can do to curb your water consumption.

First up, the toilet. It may seem a bit gross to people who haven't lived somewhere with a drought not to flush the toilet if you've only done a pee, but for those who have lived through drought conditions, the phrase 'If it's yellow, let it mellow' will be familiar. Just put the lid down.

If you don't have a dual flush on your toilet, you can get a water-saving device, or 'hippo', which takes up space in your cistern, so it uses less water to fill up and sends less water down in each flush.

While you're busy saving water with the toilet, don't undo all the good work by leaving the shower running at the same time. It's all too easy to switch on the shower, leave it for a few moments to warm up and then use those few moments to go and do something that takes much longer than the couple of seconds you actually need for the water to heat up. Instead, just hold your hand underneath the water until its warm, so you can get in as soon as it's hot enough for you. And you could get a shower timer. These can just be really simple hourglass-style timers that stick to the side of the shower screen or wall with a little suction pad. You flip it over as you get in, and when the sands have run through it's time to get out and greet the day.

A lot of us are probably cleaner than we need to be, so you could consider skipping a shower altogether now and then, especially if you're having a day at home. You could always just have a quick wash of your face, hands and a few other bits at the basin, using a flannel. Try to avoid wet wipes, which the UK uses at a rate of 3.4 billion per year, and contain plastics that damage the ocean when they're flushed. Both your skin and hair could do with a little less washing (particularly with cleaning products) and drying, so it's a win for you and for the planet.

While we're on the subject of wet wipes, be kind to your drains. In the same way that water doesn't just come from a tap, it doesn't just go down the drain. Somewhere along the way that water is going to end up back in the environment, and even in places with good sewage treatment, other things can escape into rivers or the sea, or just block up the drains before they get there. So nothing should be going down the toilet apart from toilet paper and what comes out of you. Everything else should go in a bin.

It's a good idea to get a plug for the sink, so you don't use as much water for washing as you would do if it was draining straight down the plughole. It's quite simple to fit low-flow tech on your taps too. They are just little aerators that fit on to your taps where the water comes out, and which cut the amount of water coming out significantly, without you noticing any difference to the flow. Your water provider may well give them out for free and even come round to fit them. But if not, there are lots of videos online about how to do it.

Once you've washed and you're in the kitchen, continuing your morning routine, there's more that you can do to save water. Make sure you use a plug or bowl when you're doing the washing-up in the kitchen sink, rather than just leaving the tap running the whole time. This will save water, and also means you won't need to wait for the tap to run hot, because the cold and hot water will mix in the bowl or sink. You can also fit low-flow aerators to your kitchen taps.

If you have a dishwasher, make sure it's full before you run it through and use the eco setting, as this will save both water and energy.

If you have a garden, install a water butt or three to catch rainfall to water the plants with, and make sure you only water them first thing in the morning or last thing in the evening. That way, more of the water will actually get to the plant rather than simply evaporating in the warmth of the day.

And if water meters are optional in your area, consider getting one, and you will be able to see the financial benefit of the water you're saving.

If you're clued up about water, you're also in a better position to challenge your water provider about what they're doing on a whole host of issues. If you live in a more rural area, find out what they're doing to help farmers to cut abstraction – that's the diversion of water, via ditches, pipes and streams to irrigate farm land. This reduces levels in rivers and lakes. Ask what they're doing to use water wisely in agriculture, and to cut the amount of pollution and soil that is running off fields and into rivers.

And ask if they've had any pollution incidents and what they're doing to curb them, and how they're cutting leaks, which can be a huge source of wasting water. Tell them you're doing what you can to reduce water use, but ask what their policy is for encouraging other people to do the same. Write, email or get in touch with them online or on social media. The questions that get answered are the ones that get asked, after all.

Water and wildlife

Rivers, lakes and wetlands cover just 1 per cent of the world's surface, but they are hugely important to many species of wildlife, from the flamingos that live in their thousands on lakes in Kenya, to grizzly bears feeding on the dramatic salmon runs of North America. Almost half of the world's species are found in freshwater habitats too.

But freshwater wildlife is in trouble, with populations – the number of creatures of each species – falling by more than four-fifths between 1970 and 2012, in the face of damage to and loss of the areas where they live. According to WWF's Living Planet Report, they have seen much more dramatic declines than wildlife in other types of habitat.

Sewage, run-off from farmland of pesticides, fertiliser and animal slurry, industrial pollution from industries such as textiles, and poorly planned, unsustainable dams are all damaging river and water networks around the world. We are also taking too much water out of them, for human consumption, agriculture and generating energy.

And because of climate change, temperatures of rivers and streams have risen in the past few decades, causing changes in the types of fish, insects and other creatures that are found in them.

All in all, at least 10,000 freshwater species are endangered or already extinct because of human activity. That huge loss of wildlife isn't just an abstract tragedy. Rivers flow through our lives and damage to them is damage to us, from loss of food sources to flood protection.

In the UK, the traditional dish of London's East End, jellied eels, is largely off the menu, because European eels have seen numbers fall by more than 90 per cent. The eels migrate from rivers and across the Atlantic to the Sargasso Sea before returning to breed in the rivers of Europe. The species is now considered to be critically endangered.

And in England a near-unique water system is under threat. The country is home to most of the world's 200 chalk streams, which come from underground stores and springs and flow across streambeds of flinty gravel, making them great sources of clean fresh water and a special habitat for wildlife. Species such as salmon, otters and kingfishers all live in and around England's chalk streams, along with water voles, immortalised as Ratty in Kenneth Grahame's classic children's novel *The Wind in the Willows*. But often too much water is being taken from them, and they also suffer from pollution; three-quarters of these streams are now in a bad state.

There is a growing recognition that it's not just important to preserve rivers and lakes to look after wildlife, but that wildlife that is allowed to thrive in freshwater habitats can help manage the wider landscape and deliver benefits for the environment and people.

Eurasian beavers were driven to extinction across most of their range in Europe and Asia, but reintroductions in a number of countries have helped them stage a comeback – including in the UK, where they had become extinct by the sixteenth century.

Early evidence from the schemes suggest that by building dams and controlling water flow, these ecosystem engineers can help manage watercourses to reduce pollution and soil erosion from farmland and flooding, and help store carbon.

Day Zero and the future for cities

Day Zero was to be the day in 2018 on which the taps would be switched off in Cape Town as reservoir levels dropped too low following insufficient rainfall. This drastic measure would save water in the face of a serious shortage, and residents who normally enjoyed piped water to their homes would have to queue at one of 200 collection points around the city for their daily allowance.

To avoid Day Zero – which officials warned would be 'catastrophic' for the city's economy – emergency measures demanded people cut their daily water use to just 50 litres each. That's the equivalent of one five-minute shower – but it was expected to cover each person's activities at home, work and school, from flushing the toilet to washing clothes, hands and hair, cooking and drinking.

Watering gardens, filling up swimming pools and washing cars with municipal drinking water were forbidden. Paddling pools were ruled out. Residents were urged to flush toilets with buckets of 'greywater' from the shower or their washing-up, or to follow the saying 'If it's yellow, let it mellow,' and to stick to short showers, switch off the water while shampooing, and harvest rainwater where they could. Hotels displayed signs warning about water use, and public toilets saw the taps disabled and hand sanitisers installed instead. Companies, schools, clubs and farmers all had to curb their water use significantly.

The arrival of Day Zero was at first pushed back and has since been averted, for now, after the city's residents rallied to the call to cut water consumption and the rains came. But Cape Town is not alone in facing a water crisis, and many of the world's large cities may need to reduce usage as water resources become more uncertain with climate change.

São Paolo, in Brazil, came within twenty days of running out of water at the end of 2014, while in Mexico City almost a fifth of people don't get water every day and nearly a third don't get enough. At the same time, forty-five neighbourhoods in the city face a high risk of flooding in the rainy season, and subsidence because too much groundwater has been withdrawn.

At the height of the multi-year drought in California, Los Angeles was importing most of its water from more than 300km away and the construction and refilling of swimming pools were in the firing line. In the midst of restrictions, daily use fell to as low as 132 litres a day per person on average. The drought ended with a record-breaking storm season, but after a year-long reprieve, the dry conditions returned in February 2018.

No ifs, no buts - why having a water butt can be a good thing

Anyone who has lived somewhere in the world where there's been an extended dry period will know that one of the first things to go is the garden. It's just not an option to use tap water to keep the lawn alive or the flowers blooming in their pots when water is tight for people, businesses and ➡

agriculture – not to mention the natural world. So it's a case of brown lawns, wilted roses and parched earth outside the house till the weather breaks.

But even when it isn't that dry, it's a waste to use drinking water from the tap to water plants, wash the car, or hose down the patio or drive. That water will have been treated to make it fit for humans, which takes energy and produces greenhouse gas emissions, and using it increases your water bills.

It's much better to install a water butt, rainwater tank or rain barrel if you have a garden that needs water to survive. Water butts collect water that comes off the roof into gutters and down drainpipes, diverting it for storage instead of letting it go straight down the drains. Plants prefer rainwater anyway as it is what they are designed to drink, so it's better for them to get watered from a water butt than a tap.

Water from a water butt is also good enough to clean your car, which doesn't care too much how it gets washed, and if you have a garage with a decent-size roof, you can install a water butt for that too, so the water to clean your car is literally on tap.

Climate change is predicted to lead to hotter, drier conditions in some parts of the world, and an increase in intense storms with heavy rainfall is also likely. So having a way of capturing some of the extra water we're likely to get in downpours, and saving it for the dry times, is only going to become more important. ➡

The good news is your water butt doesn't have to be an ungainly green barrel wedged up against your house. Nowadays you can get all sorts of designs, from sleek, angular, brightly coloured numbers to ones that look like old wooden beer barrels, or even giant Roman jars.

Some water butts come with attached planters with self-watering systems so that the plants not only cover up the water butt but also benefit from it with a direct water supply. And if you've inherited an old green water butt you don't like much, you can swap it for a nice new one and grow potatoes in the old one, or 'upcycle' it with a splash of colourful outdoor paint and put some trailing plants on top. Then you'll have something both pretty and practical.

Green is the new black:
Make your wardrobe last longer

Fresh from the shower, you open the wardrobe doors and scan all your different items of clothing. What to wear, what to wear? Hanging on the rack there's probably a top or a shirt that looked nice in the shop, and came at a bargain price, but really you knew it wouldn't last very long, and now the hems are unravelling and you might be thinking about throwing it away.

Or there's a dress in a colour and pattern that was very 'now' three months ago, but you fear fashion has moved on. Maybe, as you stand there inspecting your clothes, there's just too much choice, too many shirts, skirts and trousers crammed in, lots of them barely worn and unlikely to be worn again.

It's even worse at the weekends and on holiday, when there are all sorts of things to consider – especially if you work in a uniform of some kind and don't have to make decisions about dress on a working day. What's the weather going to be like? What am I doing today? Do I need to be smart or can I just slope around in casual wear? And what statement do I want to make? After all, clothes aren't just there to keep us warm and appropriately covered, they are part of a visual signal, like a bird's plumage, of who we are and what we want to say about ourselves.

But what you might not be thinking about as you weigh up what to wear is that every item of clothing has a footprint, made up of the water, energy, land and other resources used to make it, and the pollution its manufacture and even washing it causes.

Of course, when you're standing in front of the wardrobe, with the clock ticking down till you have to leave for work, there isn't a great deal you can do about the contents of your clothes collection. But you can grab a classic, well-made outfit that's in there, slip it on, and make a resolution to ensure your wardrobe is stocked with items that stand the test of time, not fly-by-night nothings.

Since 'fast fashion' became a thing, with the number of different collections a clothing chain would produce in a year rising from two at the turn of the century to as many as a dozen or more today, the amount of clothes the average consumer buys has doubled. In Europe and the US, the average person buys 16kg of new clothes a year – that's the equivalent of over two full laundry loads of washing, or thirty-two bath towels.

The clothing and textiles industry has a huge carbon footprint, using energy and creating emissions throughout the process, from growing cotton to dyeing fabrics, making garments, transporting them and selling them in shops or online. This footprint then grows even more once people have bought the clothes and they are washed, dried and ironed.

Clothes have a major water footprint too. Cotton is a very thirsty crop – while textile dyeing and the other processes fabrics and leather go through also use a lot of water – and can create serious amounts of pollution. Around a fifth of industrial water pollution comes from textile dyeing and treatment, and that's a big problem in areas with major textiles industries where water is already scarce, such as China and India.

Lots of chemicals are used in clothing production as well. Growing conventional cotton uses a significant amount of pesticides, while processing clothes with products including dyes, colourants, waterproofing materials and fire retardants all use chemicals too.

Finally, discarded clothing ends up in the bin, with four-fifths of the huge mountain of clothing the world throws away each year ending up incinerated or chucked into landfill. Some countries do pretty well on recycling, with Germany recycling 75 per cent of its clothes, but in the US it's as low as 15 per cent. Even when they are recycled, not many of the old clothes can be turned into new pieces of clothing, because the recycled fibres aren't as good quality.

So it's time to switch from fast fashion to slow, green clothes – at least in terms of their environmental footprint (unless you happen to really like the colour green). Whatever colour you go for, when you're buying clothes, it's best to spend a bit more money on fewer things that will last.

Some of our clothes are made for that: white T-shirts and blue jeans are timeless options. And you could choose something simple – a black dress or a charcoal suit never go out of fashion – and then mix it up with an unusual brooch, jewellery or tie. Or you could throw fashion out the window and go for clothes you like to wear and carry on wearing them, whatever the next fad says.

But if you want to stand out from the crowd, you can get creative. Brooches don't have to be bought, for example; you can easily make a cloth corsage, perhaps out of scraps of material or ribbons from one of those fallen-apart blouses you've got. Lots of clothes can be repurposed when they get a bit worn too. When jeans go at the knee, you can simply take a pair of scissors to them and make them into denim shorts. For the long-haired among you, a simple way to tie your hair back is with tights that you've cut into bands when they ladder.

If your children need a fancy-dress outfit for a party or school event, see if you can get away with not buying them a cheap and shiny outfit they'll grow out of in five minutes. There may be no way around it, if they're desperate for a certain look, but if they're happy to play ball, you could try to get them to be creative too. They could come up with their own superheroes and then, between you, you could create a costume from what's lurking at the back of your wardrobe.

Ordinary children's clothes can be passed on directly to friends and family, a much more cost-effective option for everybody when the little ones are growing so fast. Though you may have to find another family who is in sync with you seasonally, as you probably don't want to be dressing your one-year-old in woolly jumpers at the height of summer.

Lots of clothes end up in the bin when they could have been reused. If you're not the creative type and you're not repurposing items yourself, you can take them to a charity shop, where they could get a new lease of life with a loving owner. If you've bought stuff that lasts, then all the better, because it will go on lasting in a new owner's home. If you're not sure whether the stuff you've got is any good for resale, drop into a charity shop and have a chat with the staff there. Many shops will take pretty much anything as long as it's clean and dry, because items they can't sell will be sent abroad or used in cloth recycling, where it can be made into other things such as insulation.

And while you're in the charity shop, why not have a look around and see if there's anything there for you. Some shops have good collections of quality clothing, including last-season and end-of-line items from major retailers, so there are bargains to be had if you're happy to root around a bit.

But before you get rid of clothes, or start chopping them up for fancy dress, think about whether you could extend their lifetime in your own wardrobe. A simple way of giving your clothes a new lease of life is learning to sew. If you don't know how to sew a button back on a shirt, find someone who can teach you, and then pass the wisdom on. The same could go for hems and little holes – if you're nimble-fingered you can make an item look as good as new. You might also be on to a winner in the discount section of shops, if you're prepared to buy something slightly damaged and mend it yourself.

How you look after your clothes can have other impacts on the environment.

If you look closely at your laundry detergent container, it probably recommends you use it at 30°C, but lots of people still set their washing machine at 40°C – in fact, many washing machines have 40°C as the default for most programs. Turning your washing cycle down to 30°C won't mean your clothes are any less clean, because the laundry detergent will see to that. And you will nearly halve the energy needed to do your washing, delivering one more win for the planet and your wallet.

Another simple thing to do is make sure you put on a full laundry load, saving you more energy, as well as water. And do your clothes really need washing? Could you just air them instead, to freshen them up? If you've got any outdoor space, or even space indoors, where you can line-dry your clothes, you will make a big energy saving compared to using a tumble dryer.

A hidden impact of your laundry cycle could be the microfibres you're inadvertently washing down the drain. Clothes made of plastic, like polyester fleeces or sports tops, shed little bits of fibre during washing, which is then washed away with the rest of the dirty water and isn't fished out in sewage treatment plants. Instead it ends up in watercourses and the sea, where it joins all the other plastic building up in the natural environment. We'll get on to other things you can do about plastic later, but as a start, perhaps you could think about what your clothes are made of and whether you could avoid buying things that are contributing to the plastics problem. You could also look into getting a specially designed washing bag which you can put synthetic clothing into and then put in the machine; the bag will catch the fibres and can be emptied out into the bin after the wash has run through.

Natural fabrics have their impacts too, on land, water and pollution, so perhaps you could consider what their impact is, and if you want to try and do something about that. Look out for certified cotton which verifies the growers and the area where it's grown haven't been exposed to unsafe levels of pesticides, and that labour rights have been respected.

You could also think about which brands you're buying, and find out what their policies are on sustainable clothing. Many brands put their logos on their clothing, and if you're going to walk around advertising a brand across your chest, you might decide you want to only support ones that are taking action on this important issue.

What our clothes say about us, whether we're following fashion or wearing branded gear, going colourful or simple and sober, is a huge part of the way we dress. But just as important should be what they say *to* us. The designer William Morris said you should have nothing in your house that you do not know to be useful or believe to be beautiful. Good clothes hanging in your wardrobe or folded in your chest of drawers are both, but they can also be a third thing: items with a story to tell. Do you still own the clothes you were wearing when your partner proposed to you? Do you hold a dress and remember the long summer days on which it was worn? Does it bring to mind parties and picnics and sunsets? Is that shirt meaningful because it was the one you bought for the interview, for your new job, for the time you appeared on TV? Reducing the amount of clothes we buy and have, making sure they're things that can last, and looking after them, is not just good for the planet. It's also a different way of looking at things, allowing them to become more than just stuff that we throw away all too soon, but things of real value and meaning, part of the fabric of our lives.

The footprint of a cotton T-shirt

As you slip on your favourite simple cotton T-shirt, do you ever stop to think about the journey it's been on to get to you?

Cotton is a hugely important crop, both for the global textiles industry and the people who grow it. The fibres come from plants which put out white bolls, or protective cases, around their ➡

seeds, like balls of snow caught on shrubs after a heavy snowfall, except they thrive in warm conditions. Those fibres make up 30 per cent of the total used by the textile industry worldwide, with 100 million households in eighty countries, from the US and Australia to Pakistan and Tajikistan, involved in growing the crops.

But cotton has a big impact. It's a thirsty plant: producing your cotton T-shirt needed approximately 2,700 litres of water, around the same amount you would drink in three years. Almost three-quarters of cotton is grown on land that uses irrigation, so it's not just drinking up rainfall, it's using water taken from freshwater sources that may well be in short supply.

It also needs land to grow: around 30 million hectares (74 million acres) of land is planted with cotton around the world. And it uses more than its share of insecticides and pesticides, with all the problems they cause for wildlife, run-off into rivers, and for the health of people working in the fields. Using chemical fertilisers can also make the soil more acidic and eventually less fertile. All these things have a cost for farmers, many of whom get locked into debt cycles as a result.

Mechanically separating the cotton fibres from the seed, a process known as 'ginning', uses electricity. So that means cotton has impacts not only on water, soil health, wildlife, human health and financial wellbeing, but also on carbon emissions too, even before it starts being woven, dyed, manufactured into items of clothing, transported, sold, brought home, and later washed.

More sustainable cotton is available and makes up around 15 per cent of the total cotton grown, though only a proportion of that is actively sought out by brands. Systems such as organic, Fairtrade, the Better Cotton Initiative and Cotton made in Africa (CmiA) aim to improve the environmental and social impacts of cotton.

In Pakistan, farmers are being trained to identify which pests are harmful to cotton plants and which aren't, in order to avoid unnecessary pesticide spraying. This helps to save them money and reduces the negative impacts on their health. The ginners are being encouraged to improve the energy efficiency of their machines, saving on electricity costs, while also implementing better labour protections for workers.

In South Carolina in the US, some farmers are changing their methods to protect soil health and prevent erosion, for example by growing 'cover crops' which suppress weeds, reduce herbicide use and improve the soil.

Some brands like Ikea are leading the way on more sustainable cotton, but consumers can always ask their favourite retailers where the cotton they use comes from, and encourage them to be better.

Leather manufacturing: what have your shoes got to do with a river dolphin?

More than 400 tanneries are clustered on one river in Kanpur, Uttar Pradesh, northern India. These are the places where leather is processed from animal hide into the product that makes handbags, shoes, purses, belts and even horse-riding equipment. The industry supports 100,000 jobs in the region, and most of the leather is exported for sale in other countries.

The river the tanneries sit by is the Ganges, one of the most famous waterways in the world, and home to a rich array of wildlife, including threatened species such as the endangered Ganges river dolphin and the critically endangered gharial, a fish-eating crocodile. The Ganges is also a food source for people who catch the river's fish, and is used to water the nearby fields.

But the tanneries are polluting the Ganges with heavy metals, chemicals and other pollutants such as salt. Chromium levels in the water are as much as 100 times above the safe levels set by the World Health Organization for drinking water.

The pollution has contributed to a significant decline in the number of fish in the Ganges since the 1950s, seen most clearly at Kanpur. This is forcing local communities to change their diet, as the carp that used to make up much of what they ate has been hit more than other types of fish. And the pollution has found its way into groundwater and on to the fields where water filled with waste from the tanneries is used for irrigation.

WWF, funded by the HSBC Water Programme, is working with the leather producers and the major brands that buy the products, setting up a 'Leather Buyers Platform' to enable buyers to work together to support improved leather production practices and cut the pollution from the tanneries in order to help the people and wildlife that live in the area and depend on the river.

On the ground in Kanpur, that means working with tanneries to assess what they're doing and implement cleaner technology options. These include helping individual businesses bring in measures like the mechanical desalting of hides to reduce salt levels in waste water, and washing them in batches to improve water efficiency. It also means working at a wider scale with large groups of tanneries to better manage the waste products and reduce the use of hazardous chemicals such as chromium.

Brands that have signed up to the Leather Buyers Platform are collectively using their influence to drive change, and the tanneries engaged so far are already showing significant progress in cutting their impact on the Ganges. There are also efforts to engage with politicians on the need for more sustainable leather production across the country.

Companies involved in the scheme will also be rolling out tannery assessments in other places where they source leather, such as China. That way, it is hoped that the pollution and the environmental footprint of the shoes on people's feet around the world can be reduced.

Giving new life to a once-in-a-lifetime outfit

While it's common practice for a bridegroom to hire his wedding suit, or buy something he can wear on other occasions, the average wedding dress is likely to have been bought and is not the sort of outfit that can get a second outing. Given how much in the way of resources and money has gone into creating and buying the dress, this seems something of a waste.

Emotionally it can be a nice thing to hold on to, but in practice all too often a wedding dress just hangs in its dress bag in the corner of the wardrobe, perhaps with half an idea that the couple's daughter may one day wear it – though often they don't. Times change, styles change, ➡

and people have their own ideas about what to wear on their big day.

But a newly-wed wanting to give her dress a new life can find plenty of suggestions about ways to recycle it, from giving it to a school or local dramatics group that's in need of a wedding dress costume, to making keepsakes such as quilts or cushion covers or a christening gown from it.

There are companies that will resell dresses, so that some of the expense of buying the dress in the first place can be recouped. And charities and not-for-profit organisations will take donated second-hand or pre-loved dresses, as well as gowns donated by boutiques and designers at the end of the season when they change their stock, so they can sell them on.

Often the charities will use the money for projects that specifically support women, such as education initiatives for girls, tackling child marriage or helping the victims of sex trafficking.

And buying a wedding dress from a charity outlet can often be a lot cheaper than purchasing it from a boutique or wedding dress shop. So the bride-to-be can pick up a beautiful bargain, wear it on her special day and then donate it back to the charity for resale, so it can carry on doing good.

It's not just wedding dresses that can be given the pre-loved, re-loved treatment. Bridesmaids, who after a few years may find they've built up ➡

a collection of mostly non-reusable dresses from the times they've been there for friends and relatives, as well as pageboys, grooms and ushers, can all be fitted out from such stores.

And then there's what to wear to other big life events – a black-tie dinner, an awards evening for work. Many companies now hire out vintage or second-hand gowns and outfits, often giving money to charity as part of the deal. Often charities that sell wedding dresses also do a line in evening gowns. So instead of a once-used dress that costs a fortune and then is all dressed up in the wardrobe with nowhere to go, it's possible to look good on the big day or night, and help make fashion a bit more sustainable.

On your bike: Walk, ride or run

You grab your car keys, head out to unlock the car and jump in. It may be work, or perhaps you're doing the school run, or both, or you've got to get to the shops or pick up a friend or run any number of errands.

But it probably isn't long before you're stuck in slow-moving traffic. Your car is comfortable, but it's not so comfortable that you want to spend more time in this big metal container than you have to. And out there, in that park you can see through your window, there's a man walking a dog, and a couple of joggers, and the sun is shining and the blossom is bobbing around on the tree branches in the breeze, and it all looks so inviting, if only you had time to pull up and get out of the car. As you crawl along the road, a cyclist whizzes past.

Unless you're a long-distance commuter or a lorry driver, most of the journeys you do are probably not that far. What would it be like if you did some of those trips on foot, or by bicycle?

You might arrive at your destination energised from

exercise, not frayed from fighting traffic, and you would be getting the daily dose of activity you need to stay fit and healthy. Many of us have a very sedentary lifestyle, sitting at work all day in front of a computer, in a car, or on a bus or train to travel there, and then relaxing in front of the TV in the evening. That makes it difficult for us to get the recommended amount of daily exercise that helps us keep the weight off and helps avoid a whole raft of health problems, including heart disease and strokes.

Walking could take a bit longer, but it can be meditative. We don't give ourselves enough time to think. So walking is good for us – and it's good for the planet.

How we move people and goods around is a big part of the climate problem. Globally, transport makes up just under a quarter of the greenhouse gas emissions that are created by burning fossil fuels, according to the International Energy Agency, and the lion's share of that comes from road traffic. In some parts of the world, like the UK, transport has overtaken the power sector as the biggest source of carbon emissions.

Cars, like gas boilers in your home, are small-scale dirty power stations, burning fossil fuels to create the energy to propel the vehicles forward, making it hard to curb the emissions from each one. And combustion engines don't just generate carbon emissions, they also pump out a lot of other gases and particles – like nitrous dioxide and sulphur dioxide – that create local air pollution.

So the more journeys you and your family can make without getting in the car, the better. Often, parents take their children to school by car for logistical rather than distance reasons, so if that's the case for you or your neighbours, could you team up to walk the kids between you? You'll know how far your kids can walk, but health experts say school-age children need sixty minutes of physical activity a day, so even a slightly longer school walk could be just perfect for them. If you do drive children to school and back, make sure you don't leave your engine idling outside the school gates. It uses unnecessary fuel and causes air pollution, which evidence suggests can affect children's development.

Getting out of the car could mean getting away from the road as well. Think about how you feel when you walk through a park. Being in green spaces is increasingly recognised as being good for your wellbeing, reducing anger, stress and other negative feelings, so if walking or cycling on a designated path means you can take a route through a park, instead of along the pavement, it could improve your mental as well as physical health.

Not every journey can be done on foot or using pedal power, but if you can use public transport instead of your own vehicle if you can, that also helps. Buses and trains are a more efficient way of moving people around because they use less energy per person. And while most people know how polluting diesel buses are to their immediate surroundings (particularly if you've stood at a bus stop next to one with its engine idling), things are improving: electric buses are increasingly being rolled out in cities around the world. Even the iconic double-decker red London bus is getting the green treatment, with the first fully electric bus, which can do a full day's work on a single charge, having been introduced in 2016. There are now over seventy fully electric buses (five double-deckers and sixty-eight single-deck vehicles) in operation on the roads of Britain's capital city.

When you do drive, one simple thing you can do to reduce the carbon emissions of your journey is to practise eco-driving. And the first rule is to be more Zen about it all. Driving aggressively, speeding up and braking suddenly wears out your car, wastes fuel and just puts you in a really bad mood before you've even got to the office. If that idiot wants to cut you up or the lights are changing, let them. It'll probably make very little difference to the time your journey takes you, and smooth driving at a constant speed could deliver a significant fuel saving, so it will save you money and stress.

Observing the speed limit also delivers fuel savings, as cars become less efficient above a certain speed. It has the added advantage of not getting you in trouble! Reserving the air conditioning for when you're on the main roads going at a higher speed can also prevent unnecessary fuel use. Don't keep your boot or back seat full of junk either, because weighing down the car means it needs to use more fuel to go. And combining your trips saves significantly on fuel and carbon emissions. That's because if you do several short runs separately with the engine starting from cold each time, you can use twice as much fuel and produce twice as much greenhouse gas as a single longer trip covering the same distance but running all your various errands with a warm engine.

You could pitch in with a neighbour or relative who lives nearby to run your errands together. Even if it takes you both slightly out of your way, it will still save fuel and money compared to you both driving a similar route separately. And it's more sociable than driving around on your own shouting at the traffic and the radio. You could also see if you can find people to car-share with on a more regular basis, perhaps for the commute or the school run if you have to drive.

When it comes to buying a new car, now might be the time to start thinking about buying an electric vehicle (EV). A battery-powered EV has zero tailpipe emissions – in fact, it doesn't even have an exhaust pipe. It runs purely on electricity from the battery, most of which comes from being plugged in and charged like a mobile phone. Any pollution it creates is minimal – from the tyres and brakes – and linked to how the electricity was generated, so if there is a large amount of wind power on the grid, that's going to be 'greener' electricity than if there is lots of coal.

Until now, the main issues with EVs have been about cost, whether they will go far enough between charges and whether there will be places to charge them up again. But costs are coming down fast, and if you're buying on a car finance deal rather than just looking at the sticker price, electric cars are already comparable in price to conventional vehicles in the same class. It's also cheaper to charge up the car from the plug than to fill up a car at the petrol pump. Range – the distance electric cars can go on a battery – is also improving and so are the number of charge points available, though people who own an EV do 95 per cent of their charging at home (in general, it's cheapest to do this outside of morning and evening peaks). EVs will continue to get greener too, as more and more fossil fuels go offline, and increased levels of renewables are used instead.

A plug-in hybrid might seem a safe halfway house for now, as you can charge up this type of car to run a relatively short distance on the battery and then switch over to the petrol engine for longer distances or if it's short on charge. But because hybrids carry both a battery and a conventional combustion engine, they tend to be relatively heavy, so they can be less fuel efficient (and more expensive to run) when they're using the engine than a similar petrol-only car. And heavy cars cause more pollution from tyre friction and braking, too. So while hybrids make sense for some, fully electric vehicles are better for most.

Though the number of electric vehicles on the roads worldwide is still relatively small, it seems their time is nigh. A number of countries have set ambitions for when they want to see an end to the sale of new conventional petrol and diesel cars. The most ambitious is Costa Rica, which has announced that it wants to eliminate all fossil fuels from private transport from 2021 as part of its efforts to become a 100 per cent fossil fuel-free economy, while Norway has a 2025 date in mind. A number of countries have announced plans to aim for an end to conventional car sales by 2030, including Ireland, Israel and the Netherlands, while the UK and France have set 2040 as their date.

All these dates send a signal to manufacturers that things are changing, though the targets have to be near-term enough to be meaningful in order to really drive the shift by the industry. But car companies are responding, with increasing numbers of firms announcing new EV models or targets of their own to move away from conventional combustion engines. And the change is already happening on the roads in some places. EV market share of new vehicles is growing in many places, while China is zooming ahead with both manufacturing and use of electric cars: nearly 580,000 electric cars were sold there in 2017, a 72 per cent increase on the previous year.

It is expected that over a million will be sold there by the end of 2018.

There may be even more dramatic changes coming down the line, with suggestions that once self-driving electric cars are on our roads most of us, in cities at least, will stop owning cars and just summon a vehicle when we need it. After it has dropped us off at our destination, it will head off to pick up someone else or go back to the depot to charge its battery. If that happens, we'll stop needing all our car parks and perhaps they'll go back to being real parks for us to walk through.

Already there are many historic cities where the car isn't king. From Venice to Vienna, Brussels to Dubrovnik, Copenhagen to Cambridge, many places favour pedestrians among their narrow streets and old buildings, canals and cobbles. By keeping cars out all or part of the time, they preserve their historic qualities, and give more space for residents, tourists and wildlife to enjoy the cities.

And many modern suburbs are doing the same, such as Vauban, in Freiburg, Germany, where the 5,500 residents live in a 'reduced-car district'. There isn't any parking on quiet residential streets, so that children can play there safely, and cycle lanes and public transport links are so good many families choose to forgo a car altogether.

We may be a long way off reclaiming our streets entirely from the motor car, but if you do buy an electric vehicle soon, you may not be that far ahead of the curve. And it'll certainly be something to talk about with family, friends and work colleagues. Who knows, your decision to choose an electric car could even be the spark that starts a local plug-in revolution.

Air pollution: a dirty great problem

Air pollution is a problem that affects most people in the world, with more than nine out of every ten people across the globe living in places where the air pollution levels are above what's considered to be safe for their health.

Among the pollutants swirling around in our air are nitrogen dioxide and tiny particles known as 'particulate matter' which come from sources such as soot, nitrate fertilisers and mineral dust, and can lodge in the lungs or even enter our bloodstream. Air pollution is linked to health problems ranging from heart attacks and strokes to chest infections and even lung cancer. It has also been linked to dementia and issues with children's development.

The tiny particles in dirty air cause around 7 million early deaths a year, according to the latest figures from the World Health Organization. These deaths are caused by both poor air quality outside in the street, and in homes in poorer parts of the world where people do not have access to clean cooking facilities. ➡

Air pollution is a particular problem in low- and middle-income countries, but towns and cities in nations worldwide can be affected.

Cities in places like India have some of the dirtiest air in the world, with annual averages of particulate-matter pollution many times what the World Health Organization recommends as a safe level.* But some cities in places such as the US, UK, France, Germany and Australia also have levels above what is considered safe.

And part of the problem is transport. Research suggests cars and vans are responsible for the equivalent of a quarter of the 40,000 premature deaths each year from air pollution in the UK, and cost the NHS and society around £6 billion annually.

Nitrogen pollution from vehicles, emitted from the exhaust pipe, can upset the natural balance of nutrients in the soil, causing problems for plant species.

Other sources of air pollution include power stations and industrial plants, wood stoves and open fires in homes, and incinerating waste.

* Of the world's twenty most polluted cities, fifteen are in India. However, the Indian government recently announced a National Clean Air Plan, strengthening air quality monitoring and requiring the 100 most polluted cities to devise their own action plans to tackle the problem. ➡

To address the vehicle air pollution problem, the World Health Organization urges countries to focus on public transport in cities and making it easier to walk and cycle, as well as prioritising rail freight and passenger travel, shifting to cleaner trucks and buses and low-emissions vehicles and fuels.

There is debate over petrol versus diesel, which boils down to the fact that petrol causes fewer air pollutants, but diesel is more efficient and therefore causes fewer carbon emissions. But ultimately they are both fossil fuels and create pollution, so they will both have to go if the world is going to tackle the twin challenges of dirty air and an overheating planet.

Two wheels good, four wheels bad

Want to get on your bike, but worried about costs, safety or the weather? Have no fear – it's an easier switch to make than you might think.

Cycling has all sorts of benefits, from cutting out the need to go to the gym because you've already had a workout, to avoiding the unpleasantness of being crammed into a packed commuter train or stuck in a traffic jam. But if your commute is anything less than 10 miles in a town or a city, there's also a strong chance you'll get there more quickly by bike than by almost any other mode of transport. It's also up there with walking as one of the greenest ways of getting about.

You don't need an expensive bike, or even a bike at all, as many cities now have public bike hire schemes, either with static docking bays or a system where you can leave the bike at your destination. If you do buy one, it will be cheap compared to what you spend on public transport, filling up a car, or parking charges. It's worth spending a bit of money on a couple of locks though – a D-lock and a strong cable lock – and park up in a public place or somewhere secure, to stop it getting stolen. Cycle-to-work schemes are increasingly offered by employers, and these can significantly reduce the cost of both your bike and the necessary accessories.

It can seem daunting to get on your bike for the first time, but there are road safety courses out there, and you may be able to pair up with a colleague and cycle to work together. Testing out the route at the weekend could give you the confidence to do it too. It's unlikely to be as hilly or hard as you think it is, and in towns and cities the traffic tends to move more slowly than you expect. Cities are increasingly providing more space for cyclists, with continental Europe leading the way. If your local authority isn't a leading light on cycling, get in touch and urge them to make some changes.

If you think it's going to be too much effort, there are always e-bikes, or electric bikes. Most of the time you're doing the pedalling, but that's charging the battery so when you go up a hill and it's too much effort the motor kicks in and helps you out.

Fears you're going to spend your whole time arriving at places hot and sweaty or cold and wet may be unfounded. Regular cyclists say they get wet far less often than you might expect – and on really rainy or snowy days you can just bail. As for getting hot and sweaty, you could always shower or change when you get to work.

Shh! The future's coming: what's it like to drive an electric car?

Do you have off-street parking? Are most of your trips less than 200 miles? Well, if the answer is yes to each of these questions, a fully electric car could well be a good option for you.

EV cars are very quiet, and quite smooth as they don't have gears like a conventional combustion-engine car. But that's just the beginning of the differences between an electric vehicle (EV) and an old-style set of wheels.

A big barrier with EVs is perceptions of price, but the cost of batteries is falling, they are a lot cheaper to 'fill up', and with far fewer parts to go wrong they don't need as much maintenance.

Another big question people have when they think about getting an electric car is whether it'll go far enough before it needs charging again. That's called 'range anxiety', and it should evaporate as batteries get better and better and can go further and further, and as charging points become more widely available. Even now, you can go a fair distance on a single charge, and if you're using the car to commute, ferry the kids around or go to the shops, which for most of us make up the majority of the trips we do by car, you're unlikely ever to be out of range.

Electric car owners do report some early anxiety about range, and other issues like whether the battery will go flat if the car is left for a while. But it isn't long before owners get to know the range of their car, and start feeling confident about getting from A to B, especially if A and B are places they go between on a regular basis, such as home and work.

If that's the kind of trip people are doing in their EVs, most charging can be done at home from specially installed charge points, as long as the car is parked on a drive or off the street. When you think about it, it's more convenient just to plug your car in at home than it is to have to take it to a petrol station to fill up. Some offices, shopping areas and other public spaces are also starting to install charge points, and 96 per cent of motorway service stations now have rapid chargers – which can provide up to 80 per cent charge in as little as thirty minutes.

There's also the information the car itself is telling the driver, such as how many miles it's got left on the battery, how it's recharging when it brakes, and where charging points are, which can help the owner plan a longer journey. There are smartphone apps, too, that can put you in the picture about the location of charging points.

Other perks of an electric car can include being able to set it to start warming up half an hour before it's needed, so it's warm, demisted or de-iced before the driver needs to leave the house. If it's still plugged in, it can do all that without running down the battery.

For now, long-distance trips are going to take a bit more planning. But if you're going to the seaside for the day by car, you're going to park somewhere; you'll just need to think about where you might park and charge at the same time. There are several apps and websites that facilitate this, which are often synchronised with the vehicle for ease of use, and peace of mind. And the short wait while your car charges will give you time to sit down and enjoy a coffee and cake – much nicer than a hurried chocolate bar from a petrol station.

For now it's a novelty, and will take a certain adjustment to move over to EVs. But while for many of us it might seem a bit of a leap into the unknown, the next generation probably won't know any different from the quiet of an electric car.

Paper doesn't grow on trees: Lower your daily paper usage

Computers were supposed to make all our offices paper-free. Notepads, letters and stationery were meant to be a thing of the past, but it hasn't quite worked out that way. Yes, you get less mail in the post – instead we face an overflowing inbox. But then the phone rings, or your manager wanders over to ask you to do something, and you reach for a notepad to jot down a phone number, to take a note of a job you need to do, or simply to doodle on while you're not really paying attention.

Many of us are list makers, and a list needs a fresh piece of paper so we can be clear about what we have to do that day. Meetings can be the same. It feels so much more organised to sit down with a fresh pad of paper, ready for the thoughts and ideas thrown up by conversation to be written down.

But as you're reaching for a piece of paper at work to write notes, take a message or jot down a to-do list, look at what you're grabbing before you start writing. Do you really need another fresh sheet, or is there something lying around on your desk that would do just as well? If you've got a used piece of paper to hand, all you need to do is turn it over and you've got a fresh blank space ready to use. If post has arrived in envelopes you're not going to be able to use again, use them for writing notes and lists. If you use notebooks, make sure you write on both sides of every sheet – you can always go through them one way and then flip them over and work back through the other way if that's easiest.

To-do lists can be written down anywhere, notes can often go in the margin of an existing document, and shopping lists or team coffee orders can be written on the back of an old receipt. It may not seem much, but if you do it all the time, you'll be significantly cutting your paper use.

And paper is an obvious and tangible way of starting to think about the world's forests and how everything, including us, relies on them.

There are an estimated 3 trillion trees on Earth, and forests cover around 30 per cent of the world's land. They're one of the most important habitats on Earth, home to well over half the species on land, including three-quarters of birds. Everything from gorillas to parrots make a home in forests. Just one tree can support a thousand different species of beetles. And one hectare of forest can be crawling with 8 million ants and a million termites. That natural richness has been the source of lots of medicines, and many more could be discovered in the forests – making them mysterious treasure troves.

After oceans, forests are the world's biggest storage tanks for carbon, which trees need to live and grow. Forests worldwide take huge amounts of carbon out of the atmosphere and help regulate the climate. They're also intricately involved in our water supplies, with three-quarters of all the accessible fresh water in the world coming from rivers which are in or around forests.

Wooded areas also help reduce the risks of flooding by slowing down water, and prevent landslips and soil washing away. More than that, they help create the rain in the first place. Tropical forests store huge amounts of water, and play a big role in clouds being formed and rain falling.

They're hugely important for the people who live in them or rely on them to survive, and not just for water supplies. Around the world, 300 million people live in forests, including 60 million indigenous people. And 2 billion people directly rely on them for shelter, livelihoods, fuel for cooking and heating their homes, and food supplies, as well as water.

If you're in the middle of a city, forests may not seem so relevant to your life, but studies suggest that, like other natural spaces, they can have a significant impact on our health and how we feel. Time spent in woodlands can alleviate the physical symptoms of stress, from lowering your heart rate to reducing stress hormones, and according to studies in Japan can even boost the immune system. So whether it's a stroll through woodland carpeted with spring bluebells in Britain, or 'forest-bathing' in Japan – the practice of being in the woods and engaging all your senses while in it – time among the trees is good for us.

The fact that they're in trouble is something we need to take action to stop. Forest clearance is a major contributor to climate change: cutting down forests and trees is responsible for about a tenth of global greenhouse gas emissions. On the other hand, reducing deforestation and replanting trees could play a major role in curbing climate-warming emissions.

Paper has a role to play in that. In rainforest-rich Indonesia, for example, the leading cause of deforestation has been the establishment of oil palm and wood fibre plantations, mainly for the pulp and paper industries. Nearly 1.6 million hectares (4 million acres) and 1.5 million hectares (3.7 million acres) of primary forests – an area larger than Switzerland – were converted to oil palm and wood fibre plantations respectively. Deforestation in Indonesia is threatening the survival of species such as the orangutan.

Obviously we're not all getting our paper from Indonesia, but cutting down on paper use and making sure the paper we do use is from sustainable sources is going to have benefits around the world. And despite the technological revolution, there is probably still room for improvement.

Start by thinking before you print. Do you really need to print out that document, and if so, can you limit the number of pieces of paper you use by printing on both sides, or by checking you aren't using a sheet for a single line at the end of an email? It's not just about saving trees: it can take as much as 20 litres of water to produce just one sheet of A4 paper, and uses toxic chemicals which can pollute water courses, so reducing your paper footprint helps conserve fresh water too.

In the UK we each use an average of 145kg of paper a year, and in the US it's even higher, at 215kg a year. When you think about how light paper, tissue and even cardboard are, that's a lot of paper a year. So using both sides of the paper and not printing things out is just the tip of the iceberg when it comes to reducing how much you use.

Try to go paperless for travel and events, too. If you've received an e-ticket via your phone, resist the urge to print it out for added security, just make sure your phone is charged when you turn up to the event. If you're not sure how to find your way, why not take a photo or screenshot of a map rather than printing one out? Don't ask for receipts unless you need them either. Your bank, energy company or other services may well offer you paperless statements or bills, which is definitely a way to cut your paper use.

All these things can reduce paper use which, along with reusing all the paper you do use, can cut your contribution to the paper problem.

When you've finally finished with paper you do use,

particularly if it is good-quality office paper, make sure you put it in the recycling. If you don't have office recycling bins, ask your boss if you can get them. If you are the boss, make sure you get that sorted.

It's also something you could mention to your colleagues and team. You're much more likely to stick to efforts such as reducing paper if you tell people about it. You could even try and start up a 'page a day' campaign in the office, setting yourself a target of not using more than a sheet of paper every day, and encouraging others to join in. If you do, maybe they'll get on board with it, or even start doing other things to help the environment too.

One easy way to get the team involved and cut paper waste could be to make the coffee run more sustainable. In countries like the UK, where we use over 7 million a day, people get through billions of disposable hot-drink cups a year. And though they look like they're just made of paper, they're lined with plastic, which means they can't be recycled in general paper recycling, so most of them get thrown away. If your team has a group coffee run in the morning, turn down the cardboard multi-cup holder, and avoid the

plastic stirrers. You can deal with sugar back at the office. In fact, why not see if you can get everyone to buy a reusable cup? Or you could go the whole hog, if you've got the facilities, and ensure you have a team tea round complete with china mugs.

And maybe it's worth reminding yourself why you're doing all this. If you're lucky enough to work within walking distance of a wood or any little wild area with some trees, perhaps you could take a stroll there in your lunchbreak, to feel the benefit of walking among the trees, have some time away from your screen, and destress a little amid a busy world. The same goes for where you live: if you have one nearby, take a walk to a wood in the evening or at the weekend to reconnect with nature. You could even sign up to volunteer with a local or national organisation to help manage your local woodland or nature reserve!

Outside the office, there are other things you can do to save paper. When you're buying paper, you should aim first of all to buy recycled products, because that way you'll know that trees weren't cut down directly to make them. So whether you are buying greetings cards or printer paper, notebooks or diaries, look to see if they are made from recycled paper. At the shops, say no to paper bags as well as plastic ones, and take your own instead. You can look out for recycled toilet roll and kitchen roll too. Some people think the idea of recycled toilet paper sounds a bit gross, but obviously it's not recycled from toilet paper; instead it comes from other paper products, such as office paper.

If you can't find a recycled paper version of what you're looking for, you can look out for paper that comes from sustainable sources, such as products with the Forest Stewardship Council (FSC) mark, which ensures the paper comes from well-managed forests.

Another thing to try and tackle is the wrapping-paper mountain. We generate loads of wrapping-paper waste at times like birthdays and Christmas, a lot of which can't be recycled if it contains plastic, foil or glitter, or if lots of sticky tape has been used to secure the parcel. But it can be reused if it's looked after, and though we can all be big kids at heart, perhaps we don't need to rip away at the wrapping to get into the present. Unwrap your gifts carefully and save the paper for reuse later. Or, when you're wrapping them up, try learning the skill of wrapping with ribbon or string instead of sticky tape. That way the paper is easier to use again, or recycle if it's just paper wrapping without tape entangled in it. Look for wrapping paper or cards with great colours but no foil or glitter – and make sure they are recycled or carry the FSC logo.

You could even get creative with what you use for wrapping in the first place. Perhaps colourful adverts from a magazine or newspaper would work well for smaller gifts, or you could reuse the pages of an old wall calendar if they've got nice pictures on them. You could get the whole family involved, and set a challenge for who can come up with the best, the most wacky or the most unusual wrapping paper at a family celebration.

If you are giving a really big present, such as a bike or a playhouse for the kids, throw a colourful sheet or blanket over it and stick a big bow on top. After all, the most important bit is the present, not the presentation.

All this is good for reducing the amount of paper we use, and though we associate trees intimately with paper – and timber, which we'll get to later – it's just one piece of the puzzle of how we can help the forests that help us so much. Much of the way in which humans have an impact on the world's forests is less obvious. And one of the really big ways we have an impact on the world's forests is what we eat. It's time to talk about food.

A tree and a tick: how do you know that what you buy is sustainable?

The Forest Stewardship Council was set up in 1994 as the world's first forest certification scheme, and ensures responsible management of forests and the sustainable trade in wood and other products such as paper.

Now hundreds of millions of acres of woodlands across the world are certified under the scheme, from Scandinavia to the Congo Basin, and Brazil to China, including privately, publicly and community owned forests. Certification can be given to plantations of trees as well as to natural forests.

Wood is a renewable resource, as trees can regrow, and if produced without causing damage, wood can be a better product to use than other, non-renewable sources such as concrete or plastic made from fossil fuels. And it can also provide a valuable income for people living in forested areas.

But if not done sustainably, harvesting timber can also threaten the lives and livelihoods of those people, and put wildlife at risk of extinction, with illegal and unsustainable logging found in many parts of the world.

Under Forest Stewardship Council certification, the way forests are managed is judged against a series of agreed standards. These include the way the forest is logged, such as limiting the clear-cutting of trees (when every tree is cut down in an area), as well as restricting the use of some of the most hazardous chemicals and protecting soil from erosion. There are also standards on protecting rare wildlife and requirements on social issues, such as ensuring the rights and resources of the millions of people who depend on forests, protecting the economic well-being of workers, and making sure communities are included in decision-making.

Studies show that FSC-certified forests provide better working and living conditions and opportunities for people to engage with companies on issues related to the forest, compared to non-certified forests. They have also found that applying the standards has reduced damage to forests in certified areas compared to conventional logging, and the scheme has better measures to protect wildlife and their habitat than when companies don't have the certification.

That's even true in parts of the world where enforcement of laws governing how people look after forests is weak, and there's less support for protecting them. For example, in Gabon, west Africa, where forests are home to lowland gorillas, chimps and elephants as well as a host of other birds and mammals, companies with logging concessions which were FSC-certified were shown to be much more compliant with laws on protecting wildlife than non-certified companies.

Certification is not a universal remedy for the crisis facing the world's forests, and cannot replace strong regulations and laws for managing forests that are based on sound scientific assessments. But WWF, which helped establish the FSC, believes it can help ensure environmentally responsible, socially beneficial and economically viable management of forests.

So you can look out for the FSC certification mark – a tick which turns into a tree – on products such as paper and kitchen roll, which shows they were created with material from forests that are responsibly managed or recycled sources.

Rainforests: teeming with life

From the tropical forests of Indonesia with their orangutans and tigers to the Congo, which is home to gorillas, chimps and elephants, and the Amazon with its jaguars, anacondas and sloths, the world's rainforests are known to be rich in wildlife. Yet our knowledge of what lives in the forests, and what we might be losing before we've even found out about it, if we cut the trees down for paper and timber and to make room for agriculture, is still surprisingly limited. Though, for example, Western naturalists have been exploring the Amazon rainforest for centuries, it still remains a mystery to modern science in many ways.

In recent years alone, hundreds of previously unknown species have been discovered, from birds and mammals to plants, fish, reptiles and amphibians. They include a new primate, the fire-tailed titi monkey, with distinctive grey and red fur and a long reddish tail. As is the case for many species in the Amazon, the biggest threat to the newly found monkey is deforestation.

A pink river dolphin has also been identified as a separate species in its own right, while other creatures recently discovered include a strange-looking miniature catfish just an inch long, a snake that likes hiding buried in the soil or under tree trunks, and a yellow-moustached lizard. And ➡

there are new birds, including the western striolated puffbird, which has well-defined eyes and a strong beak and whose Latin name (*Nystalus obamai*) pays tribute to former US president Barack Obama.

The puffbird is found, among other places, in the Chico Mendes Extractive Reserve, a protected area of some 950,000 hectares (2.35 million acres) in the state of Acre in the Brazilian Amazon, where the local community has rights to harvest timber and other products such as nuts from the forest.

In December 2017, eight camera traps were put up in the reserve, which is home to 10,000 people, as part of efforts to monitor the wildlife and check what, if any, are the impacts of harvesting the timber and other products. In the first round of monitoring, 116 records of animals were made, with more than twenty species caught on the automatic cameras, including ocelots, deer, monkeys, armadillos, opossums and the grey-winged trumpeter bird.

And for the first time in the reserve, a family of pacaranas were recorded on camera. Pacaranas are slow-moving rodents found in the Amazonian rainforest, and are rare in Brazil, though more frequently found in Peru, Bolivia and Colombia. Little is known about them, and there are few records of them in the wild, especially in Brazil. Hunting and the destruction of their forest home are the major threats they face.

But conservation experts from WWF and community members say the fact that the pacaranas have been sighted on the reserve is good news. It means that as long as the right measures are applied, it's possible to use the resources the forest provides while ensuring the animals continue to survive there.

It's a long way from one to 7 billion, so start with your office team

Whether it's printing on both sides of the paper, cycling to work, turning off the computer or using the office stairs instead of the lift, there are a host of things that can help the planet and your own life, but it can often feel that you're on your own and it's easy to give up.

On the other hand, it's good to feel part of a community with people you know. Tell them what you're doing, and you are more likely to stick to your efforts, as well as getting other people involved to scale up the change. The founders of online platform Do Nation, which helps people commit to making small behavioural changes that add up to a bigger impact, recognise this.

They have come up with a system where people can make and share their pledges for greener, healthier living, and then measure their impact on carbon emissions, water use and waste.

People can create a campaign for others to sign up to. For example, instead of asking for sponsorship to do a distance swim, they can ask friends and colleagues to make pledges such as using the stairs at work instead of the lift; when pledges are made, they'll see the carbon they will save. It can even form an alternative to a traditional wedding gift list.

Do Nation's chief executive, Hermione Taylor, first road-tested the idea when cycling from London to Morocco. Instead of asking for traditional sponsorship, she asked people to donate simple actions such as cycling to work or eating less meat. As a result she managed to encourage people to save eighty-four flights' worth of carbon emissions, instead of just the two flights she saved from cycling rather than flying to Morocco.

Thousands of people have since made pledges, while companies and organisations can also use the platform to help their employees improve sustainability and their own well-being. It doesn't have to be someone with 'sustainability' in their title who launches a work scheme either – anyone can do it!

With dozens of actions on the site that people can sign up to, from lift sharing to switching off the lights, the idea is to make pledges that are doable, and will add up to make a difference. ➡

It also encourages people to stick to their pledge for two months, as that makes it more likely they'll carry on with the habit after the end of the campaign, and follows up with them at the end to see how they've done.

Do Nation believes one of the biggest results of the scheme is the conversations it sparks, and the ways in which people who have signed up to an individual pledge start thinking about how they can have a wider impact. And that can have knock-on effects that are beneficial for people and the planet.

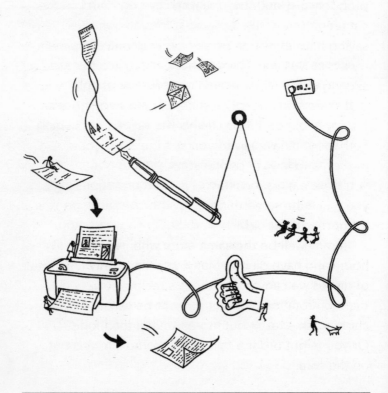

6

Gear up for a greener plate: Moderate the meat and enhance those plants

It's lunchtime and you're going to get some food. You're not feeling inspired about your meal, so you'll probably just grab the usual chicken sandwich in the canteen or from the same shop you always go to, saying hi to the same person behind the counter who always serves you. Or maybe it's just a faceless stranger serving at the till, or even a machine shouting at you that there's an unexpected item in the bagging area. It's not unexpected, you mutter to yourself, it's lunch. Perhaps you have saved yourself the hassle of going out for lunch by preparing something at home while you were eating breakfast. The only problem is that you make the same thing every morning and, frankly, it's getting a bit old.

It may well be the same story when you get home and have dinner. Many of us have only a handful of dishes we can cook without a recipe, so we eat them on rotation. Tonight it might be sausages and chips again – tasty, but more like fuel than food. Or you might order a takeaway, so you can veg out on the sofa.

But just for a moment, look at your meal, and think about where it's come from and how it could be different – how you could have created a delicious dish with a riot of vegetables, beans, lentils, spices and flavourings. How you could have swapped the same old meat for something more adventurous, without overloading on the protein you probably get too much of, while making sure you're not overstressing the planet. Are you ready to veg in, not just veg out?

There's no denying that talking about eating meat can be a difficult and divisive subject, one that provokes strong reactions from committed vegetarians, vegans and carnivores alike. But wherever you stand, hang in there for a moment, because how we eat has really significant impacts on the environment.

Worldwide, food production has a major carbon footprint. Globally, food and agriculture are linked to 20 to 30 per cent of all greenhouse gas emissions. There's methane from livestock and dairy – the infamous burping and farting cows – but there's also nitrogen emissions from the fertilisers used to boost harvests of crops. Less obvious but very serious nonetheless is the clearance of huge areas of forest and other habitats in order to turn the land over to produce a whole host of food products, whether that's farming for animal feed or for cultivating the likes of pigs and poultry. When the trees are cut down, it releases the carbon that's been stored (often for many centuries) in them and in the plants and soils beneath.

It's not just carbon release that's a problem when

land is cleared for agriculture. Those forests and ecosystems will have provided a home to some of the world's richest collections of plants and animals, many of which are threatened as a result of the loss of their natural homes. From orangutans in Borneo to the jaguar and giant armadillo of Brazil's Cerrado savannah, many species are being pushed to the brink by the growing demand for agricultural land. Though in northern Europe we famously create rich green tapestries of fields of crops, these masterpieces of food production have cost us a historic loss in biodiversity. Around three-fifths of all wildlife losses in the world are linked to agriculture.

Growing food is also thirsty work, with almost three-quarters of the water taken from streams, rivers and groundwater supplies worldwide used for agriculture, and we've already seen how that can be a problem for people and wildlife.

Obviously you can't stop eating – you'd starve. But there are things you can do to make your food footprint tread a bit more lightly on the Earth. The first thing is to change your diet so that you eat more plants and less meat.

It's way more efficient to get the protein and calories we need from veg and pulses, rather than via meat production. That's because you use up all that land, water, fertiliser and pesticide for growing crops which you then feed to animals, instead of humans, and then you need more water, land and energy to rear the animals.

The world produces millions of tonnes of soy and maize a year. Key places like the Cerrado, a woodland savannah that stores as much carbon as the Amazon, have been transformed from a natural wildlife-rich habitat into a large-scale agricultural monoculture. From a size of around 200 million hectares just twenty years ago, less than 100 million hectares of the Cerrado is left.

Most of this soy and maize is not for humans, though: three-quarters is used in animal feed, particularly for pigs and poultry and much of this in other countries like China. As a result, along with the impact of burping cows, the livestock sector as a whole is a big contributor to greenhouse gas emissions due to the land cleared to grow animal feed. As more people in more parts of the world become increasingly affluent and turn to a meat-heavy Western diet, that is only going to increase, with knock-on impacts for wildlife, habitats and the planet.

Shifting the focus of your diet towards plant-based food will not only help to reduce the environmental impacts associated with meat, but it can also benefit your health and well-being. It's not like people eating Western diets need more protein – in fact we're eating as much as double the protein that we need on a daily basis. But you don't immediately need to turn to a vegan diet, or even necessarily a vegetarian one; just simply cutting down your consumption can have a big positive impact. This should be based on an informed decision of your own individual needs – after all, everyone is different.

It is easier than you think to experiment. Lots of colourful dishes from around the world are naturally meat-free or can be easily adapted to be, from pasta dishes to a Thai green curry or a Mediterranean tagine. You can get creative and explore new cuisines, new flavours. After all, we're omnivores, we're designed to eat all sorts of things, so why not make the most of it? Imagine how boring life would be if we were pandas and all we ate was bamboo. Find a few recipes online and go for it; even if you do it for one meal or one day a week, it will make a difference, though be careful not to 'reward' yourself by going overboard on meat the next day!

When you're eating all that veg, it's good to try and eat seasonally. Do you know when broccoli is in season, when corn on the cob is at its best, or when asparagus is flourishing? Find out, and eat with the season. It makes sense that it's much tastier when it's at its fresh best – think about the flavour, or absence of flavour, in the tomatoes you get in winter compared to the ones in summer. Seasonal vegetables are also likely to be cheaper than unseasonal alternatives, and grown in a non-intensive, more natural way with the help of actual sunshine, which has a far lower environmental impact than intensive methods where we have to substitute for nature. There's also some evidence that seasonal food is more nutritious.

That doesn't mean you have to spend the winter eating swedes, and only get peas for three weeks in the summer. We think of fresh as good, and canned and frozen as bad, but often the frozen or canned version has been processed at the peak of the vegetable's freshness, preserving the nutrients and meaning you can eat it at any time of the year. And while some of those foods might be produced seasonally in a very different part of the world, don't worry too much about food miles, as transport is generally only a very small part of the overall footprint of the food that gets to your plate.

If you're thinking about buying locally and like to know where your food comes from, consider the farming systems where you live and whether you can support farmers who are producing food in ways that you feel are good for the natural world or delivering high standards of welfare. As well as certification schemes to tell you how food is produced, there could be opportunities to visit farms if you're interested. Or you could even just take a look at the countryside as you pass through it, to see whether wildflowers, wildlife and trees are thriving or if it is just field after field of the same crop.

If you're buying less meat, you're probably saving money, so that gives you the chance to support farmers producing better meat through less intensive processes, which can be more nutritious and plain tastier as a result. Eating less but eating better also helps ensure that the low-impact, high-welfare farmers who produced higher-quality, more expensive meat don't go to the wall while the large-scale, cheaper, intensive and more damaging livestock industry ploughs on.

That goes for food bought from abroad too, where many small-scale farmers rely on their crops for their livelihoods. Look for labels such as Fairtrade and Rainforest Alliance, which can be found on everyday favourites like tea, coffee, chocolate and many fruits including bananas. The labels show that farmers are getting a fair price for the work they have put in to produce your food. It addresses some of the inequalities of the global food system and often encourages the producers to implement environmental measures such as planting trees and reducing pesticide use.

If you're eating seafood, look out for the Marine Stewardship Council sustainability certification logo. That means the seafood has been assessed to ensure it is caught in a way that does not damage the environment and healthy populations of fish remain. And while you're getting adventurous with your vegetables, you could also try out some different seafood. We're creatures of habit – in the UK, for example, people mostly eat only five types of seafood, known as 'the big five' (cod, salmon, haddock, tuna and prawns). Branching out can relieve pressure on those stocks, and make for a tasty alternative dinner.

Once you've bought it, try not to waste it. Globally, a third of all food is wasted, for a whole host of reasons: poor roads and refrigeration in developing countries can mean it never gets to market; 'ugly veg', which supermarkets think their customers don't want, is often left to rot in the fields; harvests may arrive out of sync with demand; there can be confusion over best-before dates; and sometimes people simply buy too much food in one go, so it spoils and ends up in the bin.

Wasting food in the home seems to be a perennial problem, but there are lots of ways to prevent it, such as freezing leftovers, only buying what you need and learning new recipes to use up ingredients in inventive ways. You can also support 'ugly veg' aisles in supermarkets, where you have the opportunity to buy cheaper produce that doesn't look quite so perfect but is just as good for you. When you're weighing up what to buy in the supermarket, think about what that food embodies – all the time, energy, water and resources that went into growing it and getting it to you – and try to think about whether you need it, or how you can use it all and do your best to make sure it's not wasted.

Because, ultimately, this isn't about whether you're a vegetarian, a 'flexitarian' who is sometimes vegetarian and sometimes not, or even a 'demitarian' who tries to eat veggie half the time. It's about the value of food, what it takes to produce it and what we do with it. We can't individually unpick or change a vast and complex global food system, but making more conscious decisions and changing behaviours can begin to shift the dial. Our behaviour can start to raise broader questions, and highlight bigger needs than just providing calories.

Do you know what's in your fridge? Hidden soy

Soy may seem to be the preserve of vegetarians and vegans, as they tuck into another meal of scrambled tofu on toast. But there's lots of soy lurking in most people's fridges: it's just hidden in other things.

Soybeans were first cultivated in China nearly 6,000 years ago, making them one of the first food crops, and they are the 'king of beans'. Consisting of 38 per cent protein, they have a much higher protein content than meat, milk or even eggs, and they also have a more complete range of essential amino acids than most other foods, as well as a fair bit of unsaturated fat.

The US first perfected growing soy as a single large-scale crop, but it is in South America where production has boomed in the last twenty years, nearly trebling in that time, with parts of the Amazon, the Atlantic rainforest and the Cerrado converted to make way for monoculture farming. Some 113 million hectares of land are now used to grow soy globally, an area almost the size of the UK, France and Germany combined.

But the majority of the 284 million tonnes a year of soy produced on that land doesn't end up as tofu or soy milk that we can consume and benefit directly from its protein richness. That's because three-quarters of all the soy grown in the world is used to feed animals that we then consume through meat, eggs and dairy products such as milk and cheese. In a Western diet, rich in meat and dairy, the vast majority of the soy we eat is 'hidden' in these foods. In Europe, the average person eats 61kg of soy a year even if a mouthful of tofu never passes their lips.

So let's take a look in the fridge and see where the soy is. Meat contains a lot of hidden soy, with chicken and pork products being particularly big offenders. Rearing chickens actually uses more soy by weight than the amount of meat you get out at the end. So when you eat a 100g chicken breast for your dinner, 109g of soy went into it. Every egg, weighing around 55g, uses 35g of soy. So in Europe, where the average person eats 214 eggs annually, that's 7.5kg of soy to keep each person in fried eggs, egg sandwiches, quiche, omelette and frittata each year.

It's a similar story for pork, with each 50g sausage needing 17g of soy to land on the plate. And dairy has its secret soy too, with a 100g block of cheese indirectly containing 25g of soy, because the cow was fed with feed made from soybeans; there's 7g of soy in a 200ml glass of milk too.

Even salmon, if it is farmed, contains a fair amount of soy because it is used as fishmeal.

One 100g salmon steak or fillet used 59g of soy in the farming and production of the fish.

It's not something that's going away, either. If the increasing popularity of meat- and dairy-rich diets continues, so too will soy production, which is currently predicted to double by 2050. But continued soy production doesn't have to come at the expense of more forest and valuable habitat, and we as consumers can do our bit to reduce the impacts of soy production. Apart from reducing your overall consumption, think about asking retailers and chain restaurants if they are sourcing soy that is certified sustainable.

Food and climate change

Whether it is fish and chips, a chicken curry or a morning cup of coffee, some of our favourite foods and drinks are threatened by climate change.

The ingredients for dishes like chicken tikka masala come from around the world, and many production regions might be impacted by global warming. Changes in temperature and rainfall patterns can have dramatic effects on yields, and increase the likelihood and frequency of pest outbreaks. Increased extreme conditions, such as longer and tougher drought periods, can also render some areas unsuitable for further production of a particular crop – something that impacts the lives of the people whose livelihoods are tied to them.

Higher temperatures could hit chicken welfare and growth, and in South America harvests of the soy that feed the chickens are likely to be affected by rising temperatures and changes to rainfall patterns. The same problem is set to hit rice harvests in India, while warmer conditions could damage vegetables such as onions and hit tomato yields.

Fish and chips has been a national favourite for the British since it emerged in the early 1860s, sold by chippies that served it up in old newspapers to keep prices low. It's such an institution that during the First World War the government refused to ration the dish, and made sure it was available to keep people fed, and boost morale. But warmer, more acidic oceans could disrupt how the standard ingredient – cod – reproduces, and push this species further north to cooler waters, while fish such as anchovies and sea bass swim in to take their place. Thirsty potato crops could also struggle in places where conditions get drier.

In Ghana, a major cocoa-producing country that supplies some of the world's most famous chocolate brands, farmers are already having to find ways of dealing with the impacts of climate change. They are growing new, more resilient cocoa plants which can withstand longer or more unpredictable dry seasons, and planting shade trees to protect the cocoa from rising temperatures.

It's a similar story in Sumatra, Indonesia, which is known for its quality coffee. Farmers in the island's northern Gayo Highlands have reported harvests falling by up to 50 per cent in the face of unexpected rain or prolonged dry spells, which they put down to climate change. Unseasonal rain hits efforts to sun-dry the coffee beans, while warmer temperatures mean pests previously only seen at lower altitudes are on the march.

Farmers are taking steps locally to tackle the problem, distributing new, more resilient varieties of coffee plant, planting shade trees in the coffee gardens to protect their plants, and building canopies to protect drying coffee from the rain. But they warn that international efforts are needed to save the region's coffee industry – and others like them around the world – from climate change.

Food waste - what can we do?

We waste lots of food in the home, and that costs us money and the planet resources. But there are simple things you can do, so you don't have that sinking feeling as more food slides into the bin or you open the fridge and know that things will need to be thrown away.

Learn to love your freezer. You might be surprised what can be frozen – sliced bread can go in the freezer as long as it's properly wrapped up, and you can put a frozen slice straight into the toaster for toast. If you aren't going to get through all that milk you've bought before it goes off, put some in a jug and pop the rest in the freezer. If you're only an occasional user of fresh chillies, and you've ended up buying a whole bag for a single dish, the extras can go in the freezer too.

If you've made a big dish and you have leftovers, you can divide them up into single portions for another day when you can't be bothered to cook or you're home late.

If you tend to get sauces like Thai curry paste from a jar and you know you won't eat enough curry to use it all up before it goes off, you can spoon it into an ice cube tray and freeze it so you've got a whole series of 'curry cubes' which you can just pop out and into a dish. The same goes for the wine at the end of the bottle, which can be frozen in an ice tray and used for stock at a later date.

Getting the portion sizes right is another good way to avoid waste – and it's also good for your waistline, as over-catering can too easily turn into overeating. The simplest way to get to know your portion size is to weigh it out, and then find a container that matches the portion. You might have a mug that is just right for a family helping of rice, or a small bowl that perfectly accommodates the right amount of dried pasta for one.

It's worth having a well-stocked cupboard. It's something you only need to do once in a while, to make sure you've got all the spices, tinned goods, rice, pasta and pulses you need. That way, you only need to buy the fresh ingredients and the rest is on hand, ready to be cooked into a lovely meal.

Don't be too precious about bruises on your fruit and vegetables, especially if you're cooking them. Cut out bits you don't fancy the look of, and eat the rest. The inside of your stomach doesn't care what the food looks like.

And finally, know what's in your fridge and only buy what you need. If you have time you can plan your week's meals, buying what you're going to need for all the different dishes and no more. If lists aren't for you, a snap of your fridge contents on your smartphone means when you nip to the shop, you can check what you need.

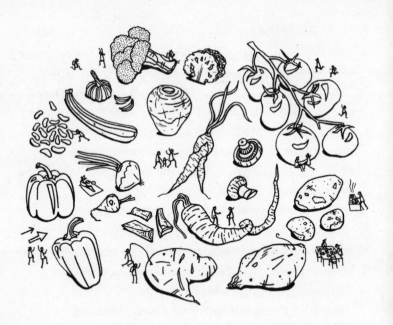

Reuse, rinse, repeat:
Cut down on daily plastic waste

Feeling thirsty when you're out and about is so unpleasant. Whether you're out shopping with the family, or heading to a business meeting, once you're thirsty and you know it, there's nothing for it but to find some water.

So you arrive at a shop, and there they are, like an oasis in the desert, those bottles of water, clear and gleaming. You reach out and scoop one up then quickly pay. You twist off the plastic cap and tip the bottle up to drink, enjoying the feel and the taste of the water, cold and fresh. A few moments later and the bottle is empty. Now what to do with it? Maybe find a bin, maybe recycle it. A few days later, you're out and about again, feeling thirsty, and soon enough you're back in a shop, reaching for another plastic water bottle.

Instead, why not get yourself a refillable water bottle, and then remember to carry it with you, so you don't have to spend money buying bottled water? Forming the habit can be the hard bit: even when people make the effort to get a water bottle, they often don't carry it with them or feel awkward asking a café to fill it up. But it's a good habit to get into, because you never know where you're going to get thirsty – in a traffic jam on a warm day, for example, or on a crowded train. This way, you can stay hydrated and tackle a major source of plastic waste – and that is something the planet really needs us to do.

The images of what plastic is doing to the world's wildlife are shocking: the albatross that flies hundreds of miles to bring home food for its ravenous chick, but all it's got in its beak is plastic; the turtle tangled in plastic fishing gear, or with a straw up its nose. When a whale that washed up in a remote Norwegian harbour had its stomach contents examined, it was full of plastic from all around the world. These individual stories are beginning to bring home to us what our love affair with plastic is doing to the world's wildlife, but they are only the tip of the iceberg.

Henderson Island, in the Pacific, is one of the most remote places on Earth. It's uninhabited and untouched, except for the 38 million pieces of plastic that litter the island, making it the most plastic-polluted place on the planet. They have been washed there from around the world by the circulating ocean currents of the Pacific.

A plastic bag has even been found in the deepest parts of the world's oceans, more than 10,000 metres down in the Mariana Trench.

Around 8 million tonnes of plastic waste end up in the ocean every year, killing and harming wildlife, and that is set to double by 2025. If things go on as they are, the plastic in the ocean could weigh more than all the fish by 2050.

Plastic is harmful to wildlife in lots of ways. Sea creatures can get tangled up in it, causing them injury or even death. They can mistake it for food, and it then blocks up their digestive tracts, fills their stomachs and starves them. It can clog up coral reefs, or be eaten by the fish that end up on our plates. It's estimated that one-third of the fish in the English Channel have plastic in them, and a quarter of fish species eaten in Indonesia and the US contain small fragments of plastic from sources such as synthetic textiles.

Because plastic doesn't decompose naturally, or turn back into natural materials, the problems it causes can persist for centuries. Over time it can break down into smaller pieces, but they're still a problem. These small fragments are joined in the natural world by plastic pieces that were deliberately manufactured to be tiny, so-called 'microbeads' which are put in face washes or cleaning products and are washed down the drain. And then there are 'nurdles', little pellets which are used to make larger plastic items. These nurdles can escape overboard from ships, or even from industrial facilities on land, and be washed into the environment and out to sea.

All these tiny plastic pieces are found throughout the seas. In the Arctic, far out of sight, each litre of sea ice contains as many as 12,000 pieces of microplastics, some far smaller than the width of a human hair, which have come from things as varied as disintegrating plastics in the Atlantic and Pacific oceans, fishing nets and ship varnish.

These tiny microplastics are often small enough to be eaten by the smallest of creatures, single-celled organisms and minuscule crustaceans, at the bottom of the food chain. They have the potential to work all the way up to larger animals and humans.

Nobody is sure what the impact of microplastics in the food chain might be, but there are concerns that polluting chemicals could attach themselves to the plastic pieces, making them toxic to absorb.

On land, too, plastic litter is not just unsightly, it can cause major problems. In India and Bangladesh, plastics have blocked drains and caused flooding in cities, while in Kenya, cows are commonly found with plastic in their stomachs at abattoirs. Plastic particles are even found in the water supplies for billions of people worldwide.

Some wildlife has adapted to put plastic waste to unusual uses. Bowerbirds, which collect items of the same colour to make a striking nest display for females, have been recorded adding bottle tops to their collections of flowers and leaves, while black kites use white plastic bags as a keep-out sign on their nests. But just like in the oceans, plastic can harm animals on land when they get tangled up in it or eat it.

All this is caused by the explosion in plastic use since the 1950s, much of it for packaging or other single uses like plastic straws. Now plastic is everywhere, and once you start noticing it, it can feel like you're drowning in it. Open your fridge and everything is wrapped in plastic. In your larder or your food cupboard it's probably the same story. And in the bathroom, all those shampoo bottles, moisturisers and shower gels mostly come in plastic bottles too.

And only 9 per cent of all that plastic gets recycled worldwide, with Europe and China leading the way.

Some countries are beginning to take action, from microbead bans in the US and UK, to 'deposit return schemes' in northern Europe which charge people a deposit on their drinks bottles that they can reclaim when they return the bottle for recycling. Many countries have brought in a charge on single-use plastic bags or just banned them outright.

A big part of the problem is single-use plastics, stuff we only use once before we have to work out how to recycle it. We only recycle around a third of our disposable single-use plastics at the moment, making it a place where it's easy to make a difference.

Plastic bottles are a good place to start, because they're one of the items that are most frequently found in beach cleans around the world, and we get through billions of them a year, with many of them littered or landfilled rather than recycled. By cutting out your purchases of them and carrying your own water bottle with you, you're doing something to turn the tide on plastic. You may feel awkward about asking to refill your bottle when you're out and about, but a bit of research will help you get to know your rights and where you can ask to fill up. In some places water fountains are beginning to make a comeback, and some cities in the UK, like Bristol, have 'refill' schemes where retailers, cafés and galleries that are happy to let you fill up your water bottle advertise the fact with a sticker in the window and on an app you can download to find your nearest friendly refill spot.

Once you've got into the habit of carrying a water bottle, there's a whole host of other things that can help curb plastic waste.

When you go to the shops, take your own reusable bag. Though we only use them for twelve minutes on average, plastic bags are a much longer-term problem if they get into the seas, because creatures such as turtles mistake them for jellyfish and swallow them instead of a nutritious meal. They're also a real problem on land, blocking drains and sewers and littering the countryside.

Reducing the number that are handed out by retailers in the first place can help prevent them getting into the environment. Try to get into the habit of not leaving the house without a carrier bag folded neatly away in your bag, briefcase or jacket pocket, so you never need to pick one up at the till. It's just a matter of adding one or two more things to the morning checklist of keys, wallet and mobile phone.

And when you're in shops, look out for ways to avoid plastic. Can you buy your vegetables loose? A lot of foods come wrapped courtesy of nature, from bananas to avocados and apples, so avoiding the ones with a second unnecessary layer of wrapping is a simple way to cut your plastic consumption. And try to buy produce that comes in more easily recyclable packaging, too. Glass is a good option, as it can be melted down and turned into new glass containers indefinitely, so if there's a choice between a glass and plastic jar or bottle for something like oil or peanut butter, go for glass. The same goes for rice and pasta: if they're available in cardboard packets, go for those rather than the plastic wrappers which are rarely recyclable. If you're trying to embrace a new-found love of cooking, you can go for ingredients that come in tins or cardboard packets, like lentils and beans, and fewer ready meals with their plastic trays and films.

You can also say no to throwaway plastics like straws, stirrers and cutlery. We use plastic cutlery on average for three minutes before we throw it away – and probably less if the fork breaks halfway through lunch – but it will remain in the environment as landfill or litter for decades. Look out for reusable alternatives for cutlery, like wooden forks and spoons, or maybe, as you pick the plastic prongs of your broken fork out of your lunch, stop and ask yourself why you're eating on the go like that – food deserves time, and you deserve time to enjoy it.

To avoid using disposable coffee cups, carry a reusable cup as a more sustainable alternative. Or give yourself time to sit in a café or coffee shop, with a nice cup of coffee in a china cup, and put away your phone, stop checking your emails, and let twenty minutes unravel while you watch the world go by.

On the subject of marking time, you may choose to celebrate your birthday with a cocktail that comes with a straw or stirrer, but these will still be going strong long after you're not celebrating birthdays any more. So tell the waiter you don't need a straw.

Then there's glitter, which while it looks pretty to us, out there in the environment is a microplastic. At the end of a party or children's art activity, it can get into the ground or be washed down the drain, where it joins all the other tiny bits of plastic working their way into remote corners of the Earth. Ditch glitter, and when you turn up for the party, don't bring a gift wrapped in the kind of wrapping paper with glittery bits that fall off as you unwrap it. That way you'll be giving another small present to the planet.

Plastics - the rise . . . and fall?

It's hard to imagine life without plastic in this day and age, but it has not been around that long. The first plastics, such as Bakelite, were developed more than 100 years ago, while polyethylene, or polythene as it's more commonly known, which is widely used for carrier bags and other packaging, was invented by accident in a lab in Northwich, England, in the 1930s. It was originally used in the military, for things such as radar cabling.

But after the Second World War, the commercial use of plastic took off. Now everything from TVs and toys to packaging that keeps food fresh, bottles that don't break like glass, synthetic clothes fibres, fencing and flower pots all contain plastic. The familiar single-use carrier bag was patented in 1965 in Sweden, soon spreading to shops across Europe, and by the 1980s it was in common use, replacing paper and cloth bags across the world.

Recent estimates suggest that since the 1950s we have produced more than 8 billion tonnes of plastic worldwide, most of which has not been recycled but has ended up in landfill or littering the world's lands and seas. Around half of all that plastic was produced since the new millennium, researchers from the University of California at Santa Barbara, the University of Georgia, and the Sea Education Association at Woods Hole, in the US, found.

But now, in some countries at least, the tide is beginning to turn against plastic.

Bangladesh banned plastic carrier bags in 2002 because they were blocking drains and sewers and causing flooding. Rwanda brought its ban in a decade ago, and travellers who bring in plastic bags have them confiscated at the airport.

Abattoirs in Kenya were commonly finding carrier bags in the stomachs of cattle, which helped prompt it to bring in its ban on plastic bags in 2017, while France outlawed single-use plastic bags in 2016, along with moves to end the use of plastic plates and cutlery by 2020.

Costa Rica is also targeting single-use plastic more widely, pledging that by 2021 it will be the first country in the world with a 'comprehensive national strategy to eliminate single-use plastics', and will replace them with products that biodegrade within six months.

Indian states such as Maharashtra and Karnataka have brought in wide-ranging plastic bans, with items from bags to plastic spoons on the forbidden list.

In many countries where you can still get single-use plastic bags, there is a charge for them to encourage you to use reusable alternatives. Places such as Ireland and the UK have seen large reductions in their use thanks to this levy, with an 85 per cent fall in usage in England after the rate of 5p per bag was introduced in October 2015.

Deposit return schemes, which put an upfront charge on items such as plastic drinks bottles or aluminium cans that can be reclaimed when the bottle is returned for recycling, have boosted recycling rates of plastic bottles in countries such as Norway and Germany to above 90 per cent.

Can we recycle plastic?

Plastic doesn't degrade in the environment like paper or other organic materials. While you might only be using it for a matter of minutes, it can linger in the landscape, or the oceans, for centuries. Even when it breaks down, it hasn't gone away; it's just there in smaller pieces.

So keeping plastic out of the environment and reusing is a must. Recycling plastic also involves significantly less energy than making plastic from virgin materials, and avoids getting more fossil fuels out of the ground.

There are lots of different types of plastic, and you can often see items such as yogurt pots marked with a number from 1 to 7 in a triangle made of arrows, along with the initials of the ⟶

type of plastic used to make it. But items can be made from more than one type of plastic, making them harder to recycle, and recyclable plastic can be contaminated by non-plastic materials.

And plastic can't be endlessly recycled as it degrades during the process. For example, a plastic drinks bottle might be recyclable between three and seven times, so while they can be remade into new plastic bottles, often they're made into other things instead, such as fleeces, toys or garden tables – even roadside kerbs.

Plastic bottles and containers are collected, sorted into their different types of plastic, and then into different colours. They're shredded, washed, melted and turned into nurdles which can then be used to make other plastic items.

It's not always financially viable to collect and recycle plastic – the black plastic trays that meat often comes in are so commonly missed by sorting machines, they're best avoided entirely. But having recycled plastic available can mean producers are not as reliant on fluctuating global oil prices, which affect the cost of making new plastic from scratch.

Manufacturers are being encouraged to make their products as recyclable as possible, to try and tackle some of these problems, as well as increasing the amount of recycled plastic they use. And in a recent surprise discovery, scientists accidentally engineered an enzyme that is good at digesting the plastic, known as PET, commonly used for bottles. The research was led by teams from the University of Portsmouth in the UK, and the US Department of Energy's National Renewable Energy Laboratory. In the process of studying a natural enzyme that allows bacteria to break down plastic as a food source, they engineered a better version of the enzyme. The discovery could provide a future solution for significantly reducing levels of plastic waste worldwide.

Plastic-free shopping

Go into a conventional supermarket, pharmacy or convenience store and you're surrounded by plastic. Sandwiches, drinks, salads and vegetables, shampoos, toothbrushes and kitchen scrubbing pads, it's all packaged in or made from plastic one way or another. But in some places, change is already afoot.

Cath Moore, based in Bristol in the UK, launched the Plastic Free Shop online in February 2018, selling a range of items from garden twine made from rare breed sheep's wool to alternative cleaning scrubbers made from loofah plants. On the site are wooden toothbrushes, beeswax food wrappers, paper sandwich bags and a kit to make plant pots from old newspaper rather than plastic.

She said she was spurred into action after her brother, a conservation photographer, returned from an expedition looking at marine plastic pollution in Indonesia. He was shocked by what he had seen, which prompted a family discussion on how to reduce plastic use.

The choice of items she sells is 'very much personal', based on the things she uses in her own home and which resonate with her and her family's lifestyle, and she says that it's not about simply replacing plastic for the sake of it, but doing simple things, which all gather momentum and change behaviour.

It's not just online shops that are offering an alternative to plastics, either. You can now find other stores that are returning to a pre-plastic world, lining up their products on shelves in dispensing jars, just like they used to do many decades ago. These stores don't stock labelled brands, just foodstuffs from spices to rice, pasta and lentils, which can be measured out and taken home in your own containers.

And it's not just retro shops that allow you to avoid plastic. Traditional street markets have long sold fruit and vegetables loose or in paper bags, though often plastic bags are used. But if food isn't pre-packaged, you can shun plastic and carry your produce home in your own bags.

Some conventional shops offer refills of certain brands of cleaning-product containers. Even ➡

supermarkets are beginning to back the 'bring your own container' approach to selling some foods, for example at deli counters or fishmongers.

Going one step further, in what is believed to be a world first, in early 2018 a plastic-free supermarket aisle was unveiled in Amsterdam. The aisle, in a new metro-sized pilot store of the supermarket chain Ekoplaza, has more than 700 products available to shoppers, including meat, rice, sauces, dairy, chocolate, cereals, fruit and vegetables.

It is also testing out new compostable bio-materials (which provide a potential alternative to plastic), as well as using traditional materials including glass, metal and cardboard, and rolling the scheme out to other stores.

8

Mindful shopping: Think big, before you buy big-ticket items

The windows of the shops on the high street that you walk down every day, or you drive past on your way to work, are filled with stuff. As well as the clothes shops, there are furniture stores with beds and wardrobes set out in inviting displays, mattresses that promise a wonderful night's sleep, sofas you could sink into. Next door, the phone shop glimmers with screens and lures you with the various deals you could get on them. The shop beyond that is an Aladdin's cave of wristwatches and jewellery.

When you're watching TV, every ad break is a parade of things you could buy, and while you're online, adverts flicker around the edges of whatever you're reading. And should you click on a link because you're thinking about a new TV for the living room, or a fridge-freezer to replace your old one, these adverts start to follow you around the internet, price tags flashing, asking to be bought.

But as the TV and the fridge pop up in the corner of yet another website, are you just thinking about the style, the quality, and the price? Or are you asking the more important questions: how much of an impact that appliance or piece of furniture has on the planet, in its production and manufacture, or in how you are going to use it?

Because everything we buy has an impact, in the energy it's taken to produce it and the energy it will take to run, the forest it came from, or the land, water and chemicals needed to make it. Thinking about that when we buy things, from food and cosmetics to sofas and clothes, enables us to make smarter, greener choices and helps us use our spending power in a more positive way.

We might already be in the habit of thinking about making good environmental decisions when it comes to whether or not to buy a plastic bottle or put chicken in tonight's curry. But often when we're making a big purchase, all those thoughts fly out of the window and it's just a question of what it will mean for our lifestyle, the look of our home, and our wallet. We don't stop to ask the other questions.

The first question should be 'Do I need it?' If buying fewer and better clothes will have a lower impact on the environment, and if less and better is also a good idea for our diet when it comes to meat, is that also going to be true for the big-ticket items? There's no denying that spending can give us a buzz: ownership of a thing makes us feel good about ourselves, and displaying our wealth can make us feel successful. These things can have a powerful hold on us. But will another TV in another room do anything except prevent the family sitting together and having a conversation? Will it drown out discussion over the dinner table, or draw the eyes away from the person you're sitting next to? Look around your home, at your tech and your furniture, your ornaments and books and all that stuff, and ask yourself how many of your things you could do without and still live an entirely comfortable life. The answer may surprise you. Are there better things to spend your hard-earned cash on, or is there something you'd rather do with your time than go shopping or spend hours doing price comparisons across different websites? It's worth asking the question before you find yourself pulling out the credit card.

Most people don't want a truly spartan lifestyle, and there are often things we do need, or which will make our lives measurably better, more enjoyable or more comfortable. But we can buy better when we buy. Take an appliance like a fridge. It will carry an energy label so you can see how efficient it is. In Europe, regulations driving levels of energy efficiency for appliances from hairdryers and vacuum cleaners to TVs and refrigerators have helped households save energy and therefore money on bills. In 2017, the UK government's advisory climate body, the Committee on Climate Change, said that demand for power had fallen by 17 per cent in a decade as a result of more efficient appliances and lighting, with significant savings for households as a result.

But you still have a choice between different

products. So when you're buying a new fridge, oven or dishwasher, check out the energy-efficiency rating, and consider buying the most efficient for your price range and preferences. That won't just cut your impact in terms of energy use and climate emissions, and it won't just help your wallet – it will send a signal to manufacturers and to governments that people want the most energy-efficient products available. This can help drive innovation in better appliances and further tightening of standards. That message can be amplified if you like what you've bought and you say so in the online review section. Mention the energy-efficiency rating and how that was a factor in your choice, and you're not only giving feedback to the company selling the appliance, you're also highlighting it as a consideration to other potential buyers who are reading the reviews, and that might get them thinking too.

If you're struggling to find what you're looking for, let the company know. If you're trying to buy a bathroom light fitting or a cooker hood and they all come with bulbs that are not LED, get in touch and ask why. Doing so will make them more likely to provide it.

Buying a new item inevitably has a footprint as a result of the energy and resources needed for its manufacture, so try to buy things that will last – like with clothing. Look into which brands have a good reputation for longevity and use recycled or certified sustainable materials, avoid things where the reviews say they broke quickly, and if you can spend a bit more to get something of good quality, that's definitely worth considering.

If you're buying something to replace an item you currently own, like a broken washing machine or an old sofa, it's also worth finding out if the company selling the new one will take your existing item away and if they will ensure it's recycled or reused.

Think about going down the reuse route yourself instead of buying new. You could browse through second-hand shops or house-clearance auctions, and if you find something that isn't quite right or is a bit tatty, get upcycling: a bit of sanding, a lick of paint, a good oil and polish, or replacing a piece of fabric, can turn an old piece of furniture into something as good as new. You can often also rent items that you don't need every day.

But when you do decide to buy new big or expensive items such as a table or a bed, or even a guitar, look out for certifications that show the materials it's made from have been sustainably sourced. It's not just paper and toilet roll that are sold with FSC certification. Wood products can be too, such as garden benches or musical instruments. Unsustainable and illegal logging of trees damages the forests in which they are found, and the wildlife in them. So looking out for the FSC logo is a way of making sure your wooden furniture or ornaments aren't costing the Earth.

And FSC-certified timber in large items like a shed can be better than buying a non-wood alternative such as fibreglass, because you know your shed has come from a sustainable, well-managed forest and at the end of its life it will break back down and be composted, unlike the fibreglass shed, which won't be going anywhere for centuries.

You can even look out for certifications for some of the most special purchases you may ever make. For wedding rings, it's possible to buy Fairtrade gold, which aims to ensure that small-scale gold miners are protected from the use of mercury in producing the gold, and from other dangerous practices, and that the environment is not polluted by mercury and waste water. It also ensures that miners, many of whom are women who also have to look after families and grow the vegetables and crops to feed them, get paid a fair price for their gold and a premium for social projects. Sustainable and ethical jewellery is niche at the moment, but that will only change if we ask jewellers where their products are coming from and trade buyers know that people care how they're sourced.

Most people would hope only to buy a wedding ring once in their life, but looking beyond the price tag to what's going on under the surface is also important for some of our most everyday purchases.

So when you are in the supermarket, don't just look at the packaging of what you are buying, look at what's in it too. One of the most commonly found products, with a huge impact on the environment, is palm oil. It's a vegetable oil that is found in half the processed products you can buy in the supermarket, from pizza bases, biscuits and ready meals to shampoo, lipstick and laundry detergent. Its rise as a global commodity has been meteoric, soaring from very low levels in the 1960s to more than 60 million tonnes being produced globally in 2016. That figure is expected to double by mid-century if demand increases as predicted in Asia and other parts of the world.

Around 85 per cent of the world's production comes from Indonesia and Malaysia, where vast swathes of rainforests have been cleared to make way for acre after acre of palm oil plantations.

While that has provided jobs, it has also damaged those communities who do not have clear legal rights over the land, and working conditions can be poor. And it has destroyed nature-rich rainforests, and released massive amounts of greenhouse gases into the atmosphere, as the forest is cut down or burned and the carbon-rich peat they grow on is drained, dried out or catches fire.

Whether it's because the orangutan has such an expressive face, or because we know it's one of our closest relatives, the fate of the red ape, or 'man of the forest' as it's also known, has become a symbol of the destruction being wreaked by palm oil. Scientists estimate 100,000 orangutans were killed in Borneo between 1999 and 2015, with logging, deforestation and large-scale plantations playing a significant part in their huge population decline. Along with them, many other species of wildlife are threatened by the march of palm oil plantations.

So take a look at your favourite products and see what's in them. With foods it can be easier to see what's in there, and you can also look out for the sustainability certification, the Roundtable on Sustainable Palm Oil (RSPO). But if you can't tell what's in the cosmetic brand you buy or the laundry detergent you always use, or if you want to know more about what the companies you buy from are doing about palm oil, get in touch, raise your concerns, and tell them what you want to see from them.

Because whatever you are buying in the shops and

online, whether it is big or small, a lipstick or a lamb chop, a guitar or a garden shed, a fridge or an electric car, how you choose to spend your money is a hugely powerful tool you can leverage to change things.

That change comes through the choice you make to reduce any negative impact of the item you're buying. And it comes as your choice encourages business leaders to drive change at a national and global level, to pull the levers of power to make products and the way they come to market more sustainable; better for people and wildlife.

It's your money. It's in your power to spend it well.

Upcycling: turn an old door into a bedhead, and other thrifty things

Upcycling sounds like one of those things that are very fashionable right now, but giving things new life by repurposing them is as old as the hills. From rag rugs made from stitching old pieces of fabric through hessian sacks to reusing old doors as a dining table, it's something that thrifty people or those on tight incomes have been doing for years.

But upcycling, taking old or discarded items or materials and refashioning them or sprucing them up into something new that you want, is also a green thing to do. First, it cuts your use of ➡️ new stuff, so there's less of a footprint than there would be if you bought a new table or chair or lamp, with all the energy and resources that would be needed to source, manufacture and transport it to you. Second, by reusing things, you keep them from being thrown away, so it cuts waste. And by being inventive you can find a new use for things that might otherwise not get recycled.

It is also an opportunity to get creative rather than just having all the same furniture, pictures and ornaments that everybody else has. That's not to say you have to be a budding artist in order to make something look nice. Just by taking a bit of care and spending a bit of time, you can make something look really good. And the sense of achievement you'll feel at what you have created far outlasts the buzz of spending money on anonymous stuff.

The best thing is, you're not alone in trying it; the internet is awash with ideas for how to upcycle your chair, your battered kitchen table, or those old glass bottles you've got. Some of it is really simple stuff, like taking a table and painting it in a combination of striking colours to turn it from something skulking in the corner of the kitchen to a central feature of the room. More ambitious people might want to get creative with old picture frames, a bit of wrapping paper and some bits and pieces they've collected to make a 3D ➡

artwork. And there are all sorts of ideas out there for smartening up old shelves and sideboards, creating unusual side tables from lobster pots, making vegetable racks from old office paper trays, and curtains from vintage handkerchiefs and flags.

Being able to see how things can be upcycled also gives you more opportunities to pick up bargains in junk shops, house auctions and from neighbours getting rid of things. Though you can find things you like exactly as they are when they're second-hand, often they're not quite what you're looking for. But if you can transform them with a coat of paint, a new fabric cover or some carefully placed wallpaper, they'll be yours.

Buying second-hand stuff and doing it up can also give you a chance to discover something of an item's story, like a newspaper lining the back of an old mirror that tells you where and when it came from. That shows it isn't just stuff – it's a part of people's lives.

Palm oil - what's the deal?

Do we save the orangutan by boycotting palm oil? Or is there something else that can be done?

For a start, from a consumer perspective, avoiding palm oil is difficult, given that it's in close to half of the packaged products we find in our supermarkets. And it isn't always easy to spot. But, as with many things, the picture isn't so black and white, and shifting away from palm oil completely, particularly at a company-wide scale, could actually do more harm than good.

So what is palm oil? It's a vegetable oil that is extracted from the fruit grown on palm trees in tropical regions. Indonesia and Malaysia make up 85 per cent of the global supply. Palm oil's popularity is down to its qualities: it stays stable at high temperatures, it doesn't smell and it contains a natural preservative, which can extend the life of products. These properties make it so versatile, and are a big part of the reason it is found in so many products – everything from cookies to pizza dough, margarine to soaps, and shampoos to lipsticks. As a crop, palm oil is a globally important vegetable oil, producing up to nine times the amount of oil per hectare of land when compared to other oil crops. So simply swapping to another vegetable oil such as soy or coconut could lead to more forests or other important habitats being cleared for production in other parts of the world. It could also cause harm to many hundreds of thousands of smallholder farmers who rely on palm oil for their livelihoods and to be lifted out of poverty.

There is a sustainable certification out there: the Roundtable on Sustainable Palm Oil

(RSPO). This involves producers, consumer groups, big buyers including global brands and retailers, and civil society organisations, who work together to develop and implement global standards for producing palm oil sustainably.

While it's not perfect, and environmental activists have found examples of deforestation or fires on plantations certified as RSPO, it's a big step in the right direction. The RSPO standard is currently undergoing one of its five-yearly reviews to ensure it better safeguards the natural world and communities that grow the palm oil. WWF believes that companies using palm oil in their products should buy from RSPO-certified producers as an absolute minimum, and use their influence to work with and support other actors in the supply chain, including the producers and the countries they work in, to improve the situation and make palm oil production more sustainable. This would help to create a palm oil industry that doesn't lead to more forest clearance and loss of valuable habitat for orangutans, elephants and many other species.

There should be no planting on peat-rich soils, which release huge amounts of carbon when they are drained and dried out to provide land for agriculture. And wildlife still needs to be able to move through the landscape, so corridors of natural habitat between plantations are important.

To make palm oil sustainable for people, producer companies need to adhere to good labour rights for those working in the fields, and no exploitation. And they need to support local communities by making sure their rights, such as land ownership and their right to be there, are respected.

It's not just about companies pushing for these things; consumers can do their bit on that front too. Check out what's in your ready meal or your moisturiser, and who made it – and get in touch with them to ask what they're doing to make sure the palm oil in their products is sustainable.

Sometimes, just one polite letter can change a company's outlook, and ultimately their behaviour on these things.

From Tanzania to clarinets: making music sustainable

The link between African hunting dogs in Tanzanian forests and the cat in Prokofiev's *Peter and the Wolf* may not be instantly obvious. But the connection is African blackwood, a tropical hardwood used to make instruments like clarinets, and efforts to make sure that the purchase of musical instruments made from the wood supports the sustainable use of resources and the livelihoods of communities who rely on them.

Tanzania's miombo woodlands are less well known than the Serengeti or the Ngorongoro crater, but the country's southern natural wilderness is rich in wildlife such as elephants and African hunting dogs. The Ruvuma River and its watershed, where the woodlands can be found, is a large landscape which is under threat from oil, gas and mineral exploration, and where illegal logging, small-scale mining and poaching are already a problem.

But a programme is under way to try and support communities who are sustainably managing and harvesting the valuable African blackwood, or mpingo, in the woodlands. WWF and local charities Mpingo Conservation and Development Initiative (MCDI) and MJUMITA are supporting the formal community forest areas.

People are trained in how to manage the woodlands sustainably, and the community forests help reduce illegal logging, according to village committees, by increasing monitoring and patrols in the areas that are at risk.

Communities need to generate an income from their forests though, and MCDI aims to create opportunities for rural villages to benefit from the woodlands, giving villagers skills and training to turn local production into commercial activities ➡

to improve their lives and encourage them to keep looking after the forests.

The organisation has been working for more than ten years to develop ways in which community-controlled forest enterprises can access domestic and international markets to sell their sustainable hardwood timber. This includes getting Forest Stewardship Council certification for the hardwood, which could be used to make furniture, musical instruments and flooring.

There are challenges to overcome, including making premiums for certified products a reality, finding markets, and getting the products out of the remote area, which currently has limited infrastructure, although it is improving.

Social enterprise Sound and Fair is helping to find a market for certified and ethical blackwood from the area. It has previously supplied musical instrument manufacturers in the EU and the US, including makers of clarinets, oboes and bagpipes, and is developing a local sawmill for processing African blackwood and other hardwood species from the forests, which can also be used for other items such as guitars.

The revenue goes back to the villagers, who can spend it on the things they want and need, such as building schools and water wells, and providing healthcare.

The idea is that because they're benefiting from the sustainably managed forests and

see the value of them, they will conserve them.
And people playing instruments thousands of
miles away can know they're making music without
inadvertently destroying important wild areas and
landscapes.

**Potted potential:
Clean the air by potting
plants at home and the office**

You're home after a long day at the office. As you
walk up your drive to the door, in the front garden,
or on your doorstep, on your balcony or windowsill,
there is a planter of flowers and they are blooming with
all their might, flower heads bobbing in the breeze, a
shock of colour or a calming tone against the paving or
the wall of your home. In a nearby tree, a bird sings,
staking out its home in a riot of musical notes.

It's great to see those flowers bloom, because you
spent a lot of time on that pot, choosing it, growing
the plants from seeds or buying them as seedlings in
the hope they would grow. You potted them up and
you water them when it's dry, maybe giving them sticks
and string to hold them in place. Now they're not just
blooming for you, but bugs are also able to enjoy
them: a spider spins a web whose strands catch the
light of the evening sun; a bee or a hoverfly lands on a
flower with a little hum. And that singing bird – well, the
bugs are probably its dinner.

When it rains, a little bit of the water that falls down
on the concrete, the driveway, the pavement, is caught

by the pot, taken in by the plants, stopped from dashing down the drain. In other places there are puddles, some that get deep quite quickly if it rains hard and the drains are blocked or there's nowhere for them to soak away. But the pot is a tiny oasis, where the leaves glisten green under the raindrops. The city breathes a little easier.

We tend to think of our homes and gardens or balconies as a haven, a little private island in the sea of the world, where we can escape to, shut the door on the rest of humanity and just be ourselves, live our lives. But unless we live alone on a remote island or in the middle of the wilderness or countryside, we're not as separate from each other as we'd like to think. Some of the things you can do at home, like saving energy and water or changing what you eat, help on an individual level and can build up to a much bigger contribution if lots of people do them. But there's also other stuff you can do at home that will help your immediate surroundings, and perhaps make your life a bit better too.

Take that pot on your front doorstep or driveway. If you're in an urban area and you have a garden, or a front yard area, anything you can do to make that area greener will help your immediate surroundings. Greenery can help control urban temperatures. Towns and cities are particularly at risk of heating up because of the dark surfaces such as roads and paving which absorb more heat than areas with plants and grass. This contributes to the urban heat island effect and can make heatwaves worse in towns, something which will get more severe as global temperatures increase, making life uncomfortable and even putting lives at risk. The Europe-wide heatwave of 2003 is estimated to have led to tens of thousands of deaths, with France particularly badly hit. Trees and other planting help cool buildings and streets, and reduce carbon dioxide. Trees, hedges and creepers on the house can even prevent homes from getting cold in winter by acting as a windbreak.

Green areas in cities can also reduce flash-flooding, which can be caused by large amounts of rainfall in a short period that has nowhere to go. Drains are overwhelmed, because the rain can't soak into the ground in largely paved or tarmacked areas. Trees and shrubs can briefly hold water on their leaves, slowing its fall to the ground, while earth planted with greenery helps water soak into the soil, stopping it pouring off and sitting on surfaces where it can quickly turn into floods.

Many trees and plants can also help with local air quality, absorbing pollution, not to mention noise, and acting like the lungs of the city. And, of course, wildlife benefits from more planting and less paving in your front or back garden, whether it's birds wanting to pluck worms out of the ground or bugs, bees and other creatures needing somewhere to hide and nest.

Although we often think that nature is something separate from us, something out there, having plants and wildlife on our doorstep is good for us. We've already seen how time in natural spaces like woods can improve our well-being, and lower our stress levels, and that's also true in towns and cities. A survey for Britain's Royal Horticultural Society found that almost three-quarters of people (73 per cent) believed that planted areas in roads and streets would make them feel happier, three-fifths said they'd feel healthier, and almost as many said they would feel calmer. And yet in the UK more than a quarter of front gardens are completely paved over, three times as many as a decade ago, and 5 million have no plants growing at all.

While front gardens are driveways for many people, you need less area than you think to park your car – after all, you only need two tracks where the tyres go. The rest of it could be covered with low-growing plants, or you could be relaxed about weeds in your gravel. And even if your front garden or backyard has a large expanse of grass, there's probably more you could do to create a more beneficial habitat for wildlife. Wildlife likes the scruffy places, the fallen leaves, the woodpile, the wild flowers which grant insects easy access to pollen and nectar, the shrubs and trees. So relax, your garden doesn't need to look like something out of a glossy magazine. If anybody looks like they're judging you, say you're gardening for wildlife.

Try to grow plants that will attract wildlife, for example those which attract pollinating insects such as bees, which are in decline in many parts of the world. In 2017, German scientists found that numbers of flying insects were down 75 per cent in three decades in protected areas, meaning populations could be in an even worse state in ordinary urban or agricultural areas.

Gardens can provide a haven for wildlife which has been pushed out of farmed areas. One of the things affecting insects is pesticides, so the more you can avoid chemicals the better, even if that means you get a few pests. There are other ways of getting rid of pests like slugs and snails, from beer traps to coffee grounds, which you could look into if you can't bear your prized plants being chewed. Or you could encourage other wildlife such as birds into your garden and hope they'll snack on your pesky pests. Your garden could be a haven for hedgehogs, which are doing better in urban areas in the UK than in the countryside, and will eat up your slugs – just make sure there's a hole in your fence so they can get in and out. And choose alternatives to peat compost, because you don't want to be making a lovely habitat or beautiful space in your garden while using a product that has come from digging up peatland, an important habitat which stores carbon and supports wildlife.

Whether you have a large garden or just a window box or a sunny windowsill, you can also try growing your own herbs and veg, and even fruit if you've got room. If you're trying to eat seasonally and value your food, it doesn't come much more seasonal than when it ripens in your own backyard. And it can take up less space than you think. Most sunny windowsills have room for a basil or chilli plant, and you only need room for a sack or a pot to nurture some potatoes or tomatoes. It takes time to produce your own vegetables, but there's nothing more satisfying when it comes to food than eating something you've grown yourself – not only will it be tastier because it's so fresh, but you'll know exactly how it's been produced, and the whole journey from 'plant to plate' will just be a few metres.

You can even use any food waste you generate, alongside grass cuttings and other plant clippings, to make your own compost for your garden. Making good compost is a bit of an art, but it's very satisfying to use your own compost to spread around your garden or in your pots.

It's not just food waste you can recycle, either. In the house, you'll probably be generating a fair amount of waste, so set up your bins to make it as easy to put things in the recycling as in the rubbish bin, and it becomes automatic. Make sure you know exactly what your local authority will collect or where you have to take your recycling, and how best to recycle it. For example, you may have to take the pump sprays off household cleaner bottles before you put them in the recycling. The more recyclable material the local authority can collect and make money from, the better for everyone's bills, not just the planet. It also reduces the amount of waste being incinerated or sent to landfill.

You can also make sure anything you can't recycle is disposed of responsibly. Fly-tipping is a major problem in many places, and a lot of it is simply unwanted stuff from households. It's important to make sure that anyone you get to take away your unwanted items, from furniture and appliances to rubble and old fencing, is going to take it to an appropriate place and dispose of it legally and responsibly. Make sure you get any paperwork you might need, too.

That way, you can get back to your home and garden and spend time making them beautiful and beneficial, without worrying that you've inadvertently blighted another part of your community with illegal waste. And time is something you need when it comes to nature. The poet William Blake said, 'To create a little flower is the labour of ages.' It's true, growing things takes time. But for you, and your surroundings, it's time well spent.

Green in the heart of the city

The last freight train ran on the elevated line through New York's West Side in 1980, and as the line fell into disrepair there were calls for it to be demolished. But nature had returned and, inspired by its wild beauty, Joshua David and Robert Hammond founded Friends of the High Line to call for it to be preserved as a public park.

The first section of the High Line opened in 2009, and now a 1.5-mile 'greenway' weaves between New York's city blocks. The route along the old freight line is planted with hundreds of species of trees, plants and shrubs, from dogwoods and hollies to magnolias, grasses and flowers, chosen for their hardiness and sustainability and with a focus on native species.

The High Line creates a linear park through Manhattan, a place for walking, sitting, watching performances, viewing art, getting some ➡

food and even having a dance. It's an example to other cities of how to restore old industrial sites to create a slice of greener city living.

In recent years, London has also been creating new slivers of greenery among the buildings, in the form of more than 100 new 'pocket parks': patches of land that have been transformed into new spaces, with everything from a rain garden at the dead end of a road to tiny plots by bus stops springing to life.

Recent larger-scale green projects have seen the opening up of the city's reservoirs as public nature reserves, including Walthamstow Wetlands in the north-east of the city, which is home to the UK's largest colony of cormorants away from the coast.

Now London – which has almost as many trees as people, is home to 14,000 species of wildlife, and has green space accounting for almost half the total land area – is set to declare itself the world's first National Park City in 2019, to help make it greener, wilder and healthier.

In Milan, a radical approach has seen greenery taken to new heights. The Bosco Verticale, or Vertical Forest, the brainchild of architects Stefano Boeri Architetti, is formed of two residential towers in the Italian city. They are tower blocks with a difference, containing hundreds of trees on planted terraces that run up the side of the building, along with thousands of shrubs and plants that spill over the side of the balconies.

The idea behind the scheme, which has the effect of creating living tower blocks swathed in green, is to create a habitat for birds and insects, absorb carbon dioxide and dust pollution, and provide shade and protection from noise pollution for residents.

It's an idea which is set to be scaled up to city level, with Liuzhou Municipal Urban Planning in China giving the go-ahead for Liuzhou Forest City, a home for 30,000 people where apartments, hotels, schools and offices will all be planted with thousands of trees and a million plants, to improve air quality and wildlife.

Smart homes now and in the future

With so much of our impact on the planet happening right at home, in our houses, driveways and gardens, smart homes are set to truly transform our lives.

Around the outside, homes will not necessarily seem high tech, but we'll need to take 'smart' steps to cope with our changing world. Those could include green roofs, to help absorb more extreme rainfall and to cool urban areas.

On the ground there will be permeable paving and areas with small 'swales' or 'rills', planted depressions or channels to slow the flow of water run-off. And the drive or pathway lighting could all be solar lamps.

There could also be solar panels on the house, but clunky ones bolted on to the roof are already a bit retro. Now panels sit flush with the tiles, or you can replace them entirely with solar tiles that come in slate or even terracotta styles.

The panels are one of a trio of technologies that will form the heart of a smart energy home, along with a home battery pack about the size of a boiler, and an electric vehicle in the drive. That way the house is generating electricity from the panels and storing what isn't needed by charging up the battery and the car.

If it's a new-build home, it's likely to be highly efficient, with insulated walls and roof and triple-glazed windows to keep the desired heat in – or unnecessary heat out. Homes built to 'passive house' standards use very little energy to keep them warm, are super-insulated, use the power of the sun to do much of the work, and have a ventilation system that recovers and recycles the heat from the air that is being let out of the house.

Smart technology is also likely to let us be much more in control of how we heat and light our homes in the future. It's already possible to get systems where you can turn on the heating as you head home from work, and set time preferences for each individual room in the house – so you can keep bedrooms cooler and living rooms warmer. Automatic technology can turn lights on and off as you enter and leave rooms, and could help waste less water and energy, for example with taps that turn off when you remove your hands from the flow to soap them, or showers that limit flow until the water is hot enough for you to shower.

Remote technology, which allows you to switch on your washing machine or dishwasher from afar with an app on your phone like the one for heating systems, could save you money. Higher prices for peak electricity could encourage you to load up the machine before you head to work and then set it going in the day when demand and prices are lower.

In the UK, WWF has worked with National Grid and Environmental Defense Fund to produce data that can be used for apps that will tell people in advance whether there is going to be a lot of renewable energy on the grid. For example, if it's going to be a windy day, they can set appliances running to make best use of that.

Appliances could also be linked up to the wider grid, via smart meters, which could allow appliances to temporarily power down for a few minutes, to help balance peaks in demand. So when it's half-time in a sports match, or the end of the royal wedding, and everybody gets up to make a cup of tea or open the fridge for a cold drink, the power surge is managed by powering down freezers for a few minutes, without damaging food but helping balance supply and demand.

What becomes of my old glass bottle . . . and the rest?

Recycling has become significantly more widespread and comprehensive in recent years, and many of the items commonly used in the home are now recyclable.

It's still better to reduce the amount of stuff you get through where you can, first, and then to reuse items; for example, after you've finished eating the jam in a jar, you can wash it out and use it as a container for something else, instead of buying new containers. But it's hard not to get through lots of bottles, tins and other containers, so being able to recycle them is a way of cutting the impact of all that stuff.

Though there are limits to how much plastic can be recycled, it's still better to recycle whatever you are using, rather than bin it. It's the same with fabrics, which can be turned into ➡

everything from mattress stuffing to insulation. Card and paper, including newspapers, magazines, junk mail, flyers and envelopes are turned into new paper and cardboard products.

Some things can be recycled indefinitely. Aluminium cans, foil and deodorant cans can be endlessly recycled, and doing so uses just 5 per cent of the energy and creates only 5 per cent of the greenhouse gas emissions that sourcing new aluminium for the cans would require. According to the Recycle Now campaign, recycling just one drink can saves enough energy to power a TV for four hours.

It's the same story with steel, which is commonly used for food tins. During recycling, steel and aluminium are separated by magnets that only pick up the steel. Both products are shredded, heated and melted, then can be made into new cans and tins.

Glass is also endlessly recyclable, with clear food jars, green wine bottles and brown glass all getting the recycling treatment. They can be sorted into different colours, but if they get mixed up, for example because they get broken during collection and transport from homes, then the glass ends up as brown. That's perfect for beer bottles.

Other things you might not think of as recyclable include electrical and electronic goods, which can contain metals such as zinc, platinum, ➡

palladium and even gold in things like mobile phones. The reclaimed metal can be used for everything, from preventing ships from rusting to making jewellery. The plastic from old electronics and items like lawn mowers can also be sorted and reused for things ranging from car bumpers to musical instruments. Materials from batteries can also be recovered and reused. And an old iron that's no longer ironing your clothes has enough steel in it to produce thirteen steel cans.

SMALL ACT

10

Eco-footprints in the sand: Lower your holiday footprint

It's a key part of the holiday ritual: check you've got your passport if you're going abroad, pick up your suitcase, lock up the house, and bundle into a taxi, the car or on to a train to get to the airport.

There you have to navigate check-in, often trying to skirt round children who've been allowed to push unwieldy luggage trolleys. Then there are the liquid limits in hand baggage, the taking off of shoes and watches, belts and coats to go through security, then trying to work out where your boarding gate is, and whether you need to get on a little train to reach it.

Once you get to the boarding gate, there's another queue, and then you're finally walking down the ramp to the aircraft . . . or perhaps having to get on a bus to be driven along the runway and out to your plane.

On board, there's the scuffle for overhead locker space, people are blocking up the gangway, getting in and out of seats.

Eventually everybody gets themselves sorted, the cabin staff perform the safety demonstration, the captain tells you the flight time and the plane taxis to the end of the runway. You're finally off on holiday.

But flying is one of the most climate-polluting things a person can do. So maybe it's time to think about holidays that don't cost the Earth, by swapping the flights for a different mode of travel – trains or ferries, or a combination of both, say – or changing the destination so you don't need to get on a plane. There are all sorts of upsides to a different approach to travel.

Even taking a single flight can significantly push up your annual carbon footprint. Globally, aviation accounts for about 2 per cent of the world's carbon emissions, which doesn't sound a lot, but there are several things to remember about that. First, the figure only includes carbon dioxide, not all the other emissions that planes put out and which have impacts on the climate, such as water vapour, which is also a greenhouse gas. Once you factor those in, flying has at least twice as much impact as what's caused by the carbon dioxide alone. And as the world takes steps to cut its emissions in other areas, such as energy and road transport, aviation's share of the problem is set to increase. Finally, if we're going to meet the goals in the Paris Agreement of curbing temperature rises to well below 2°C, then we've got to cut out pretty much all our emissions.

International aviation isn't directly covered by the Paris Agreement on tackling climate change, but a deal has been done by countries under the umbrella of the

International Civil Aviation Organisation (ICAO) to address the problem. It's essentially an 'offsetting' deal that seeks to cap emissions from international flights at 2020 levels.

That means airlines will have to buy permits that deliver reductions in pollution elsewhere to cover emissions that are above 2020 levels, for example by funding wind farms. The cost of buying the permits is likely to filter down to passengers.

There are some efforts under way to make air travel more 'low carbon'. Aviation fuel, which comes from oil, can be replaced by fuels that are made from plants or waste, and some of them can offer carbon savings compared to conventional fuel. But the problem is that the production of these plant-based biofuels can have all sorts of other negative impacts on the climate that mean you might not be solving the problem, but potentially making it worse. For instance, if an airline were to use biofuel made from palm oil that had been grown on plantations created by cutting down rainforests and drying out peat soils, then all that has happened is the carbon problem has shifted somewhere else.

Another alternative being explored is electric aircraft, which would see planes following the lead of cars and swapping jet fuel for a battery pack for short-haul flights. These flights would then be green if the electricity used to charge the batteries came from low-carbon sources such as wind power. US company Wright Electric is working with several airlines around the world to develop commercial planes which can be powered by batteries. These aircraft could be in the skies carrying passengers within a decade. But they don't look set to replace long-haul flights, and there are already pretty good alternatives to short-haul flights, most of the time: rail or ferries.

Unless and until we get a big technological breakthrough – a plane that can carry hundreds of passengers to the other side of the world without generating any climate emissions – the best way to cut the carbon from flying is simply to fly less. And if you're going on holiday, think of different ways to get there.

It isn't necessarily as hard as it might seem. Even if they're not going very far, most people would probably opt for a flight because it seems the quickest and most straightforward way of getting to where you want to go. But often, once you consider all the time you take getting to the airport, which is often miles out of town, waiting for your flight, your baggage and your onward travel, it practically takes a day to get anywhere. So when you think, 'Oh, I couldn't possibly go by train, it would take hours and hours,' remember that, unlike airports, train stations are usually in the city centre, so door-to-door it might even be quicker than flying.

Some places take much longer to get to if you don't fly, but you can think about the journey as part of the holiday, and embrace slow travel. We lead such busy, hectic lives, that it always takes us a few days into a holiday to unwind anyway, so why not give yourself that time while you travel? Spending hours on a train or a boat watching the world changing out of the window or from the deck gives you time to relax. It can even build the anticipation, and give you a greater sense of adventure, than just arriving, slightly hassled, at an airport that looks like any other. Or it can simply give you enough time to really get into that book you've been meaning to read.

You could even turn your whole holiday into a journey, travelling from place to place. Walk a long-distance trail, cycle from town to town, or Interrail across a continent.

If you do fly, it is possible to offset your carbon emissions, and the best option there is to go for credits certified by the Gold Standard scheme, which ensures the money you pay is delivering the actual emissions cuts to compensate for the emissions you caused by flying. The Gold Standard scheme is not just about cutting emissions; it also aims to improve people's lives, and makes sure that the projects your offsetting pays for don't have any negative social or environmental impacts. But not flying is better, if you can manage that – a tonne of carbon avoided is better than a tonne of carbon that is offset.

The kind of holiday you choose can also be a way of helping the planet. Have you ever thought about going on a volunteering holiday where you take part in nature-friendly activities? You don't have to go to a remote rainforest to help nature, there may well be things much nearer to home you could do. A holiday is meant to be a time to relax, destress and enjoy your surroundings. And many of the benefits to your well-being – reducing stress levels, for example – that you get from being in a woodland, park or garden are the same wherever you are in the natural world. If you don't have that much money to splash out on your holiday and you don't want to go too far afield, a volunteering holiday may well also be quite a cheap option.

Even if you don't choose a nature-focused holiday, there's no need to forget about the planet while you're away. You can still do all the things you would do at home, from switching off the lights when you're not in the room to not wasting water. In some places, those things might be even more important than they are at home. If you're somewhere hot and dry, water is likely to be scarce, for example.

In many parts of the world, tourism is really important to local communities, but it can have negative impacts too, from overstretching resources like water to low pay for staff. So think about trying to stay in places where your money is best going to reach local people and which don't have too big a footprint. There may be eco-lodges, or homestay options where you stay with a local family, and they might be much more rewarding than a big impersonal hotel.

Tourism activities such as wildlife-watching can also be a boon for local communities and threatened species. Visitors coming to see wildlife can be important in providing jobs and income for people living in that area, which means they benefit from conserving the wildlife. That's very important for making sure efforts to protect at-risk creatures are sustained in the long run. For example, a mountain gorilla that people can go and see can indirectly generate around £2.5 million (US$3.3 million) during its lifetime from tourist income, and that money has helped transform government attitudes towards conservation in Rwanda and Uganda. But there are still risks to wildlife from interaction with humans; if tourist interactions are not well managed, mountain gorillas, for instance, can pick up diseases from people, or the presence of visitors can interrupt their natural behaviour. If you're planning to go wildlife-watching, try to do some research beforehand so you're sure your activities are helping, not harming, the creatures you want to go and see.

And be careful about buying any souvenirs that are or look like they might be made of animals or plants. Countries try to protect rare and threatened species by preventing or limiting international trade in them or their body parts under a treaty called the Convention on International Trade in Endangered Species of Wild Fauna and Flora, which is a bit of mouthful so it tends to be referred to as CITES.

Countries also have their own national laws which can be even stricter. UK law means it is illegal to bring endangered animals or plants, or goods made from them, into the UK even if they were openly for sale overseas, unless you have a valid CITES permit. This includes ivory, shells, jaguar teeth, reptile skins, caviar and traditional medicines which may contain animal parts like tiger bone. Many other species are also protected that you may not realise are endangered, such as orchids, cacti, and even some types of wood.

It's important to ask what products are made of, where they came from, whether their sale and export is legal in the country you're in, and whether you'll need a permit to bring them home. If in doubt, walk away and buy a postcard instead. Where you spend your money can have a big impact on some of the world's most at-risk wildlife.

All aboard! World train travel

Many people have heard of the Orient Express, and Canada's Rocky Mountaineer, as they're some of the most famous train journeys in the world. But how many people know that it's easy and affordable to get from London to Italy in a day by train, with lunch in Paris and dinner in Turin?

Or, if you have flown to far-off places, once you're there it's not too difficult to take a train to get around instead of adding to your carbon footprint with short-haul flights. And by staying on the ground, you can really see something of the country you're visiting from trains, buses and ferries, and maybe even meet some local people and get to share your lunch with them.

Navigating train systems, ticket offices and station food stalls in different parts of the world can also be part of the adventure of a holiday, in a way that taking a cab to the airport and queuing up with all the other tourists never is.

British career railwayman Mark Smith founded the Seat 61 website in 2001 when he saw that people needed information about how they could swap a plane journey for a rail trip. He finds people most commonly search for how they can get from the UK to Italy and back, followed by other top holiday destinations such as Spain. As well as cutting carbon from travel, he thinks rail trips can be part of the experience of the holiday, not an interruption of it. A train journey can even be a highlight in its own right, and can help people get out of the 'tourist bubble', he says.

Some great train journeys in Europe include London to Fort William, with its stunning views of the Scottish Highlands; the narrow-gauge Bernina Express, with panoramic viewing carriages of the Swiss Alps which can be enjoyed in winter or summer; the trip from Belgrade to Bar through the remote and rugged mountains of Montenegro; and the slow route down the River Rhine.

In Vietnam, travelling from Hanoi to Saigon or Danang by train takes you along the country's coast and through towns and countryside. You can travel the breadth of Russia on the Trans-Siberian railway, while travelling by train in India is a real part of the experience for the foreign visitor.

Visitors to New Zealand can travel from Auckland to Wellington in the North Island or across the Southern Alps from Christchurch to Greymouth. In North America you can travel across both the US and Canada by train.

Trips like these can give you views and vistas and people to talk to, and often comfortable seats and buffet cars and no jet lag. All this, and you don't have the hassle of airport security.

World heritage at risk

It's a beautiful world, but some of the most special places on the planet are at risk. That includes our own built heritage, with many World Heritage Sites – such as the canal-crossed medieval city of Venice, Neolithic remains on the remote Scottish islands of Orkney, and New York's Statue of Liberty – under threat from climate change. Rising sea levels and increasing storm surges put places like these at increasing risk of damage and destruction.

And some of our best natural sites are at risk, not just from climate change but from industrial activities on the ground, such as logging, mining and oil drilling. Even tourism, when not managed well, can be harmful to the very places people want to visit.

Natural World Heritage Sites are not just beautiful places to go and look at, they are home to people who rely on them for food, fuel and other resources, they can provide clean water and flood protection and they are also a source of employment through tourism and recreation.

But many are under threat, such as the Tropical Rainforest Heritage of Sumatra, whose forests are home to a rich variety of wildlife such as the Sumatran orangutan and tigers, but which faces road development that will facilitate illegal logging and small-scale mining.

In Madagascar, the Rainforests of the Atsinanana, made up of six national parks along the east of the island, where rare and threatened species including lemurs are found, are at risk from illegal logging and forest clearance.

And Virunga National Park, in the Democratic Republic of the Congo, whose landscapes range from swamps to volcano peaks and which is home to mountain gorillas and other wildlife including 20,000 hippos, remains at risk from oil exploration.

There are success stories for natural World Heritage Sites which show how they can be managed to support local communities and wildlife, and provide a tourist draw.

The Tubbataha Reefs in the Philippines were being harmed by damaging practices, such as dynamite fishing, but the government stepped in to establish a no-take zone. This has led to a recovery of fish stocks both in and around the site, boosting catches in the surrounding area and encouraging more sustainable fishing practices. It has also trebled the number of tourists visiting the reefs, providing investment in the local area.

In Nepal, the creation of Chitwan National Park in 1973 caused conflict with local communities who were forced out and resettled outside, losing access to land and forest resources they relied on. But in 1996 a buffer zone was created which is now home to 300,000 people, and park officials and residents work together to manage the natural resources in the buffer zone. As part of this, forests were handed over to communities to manage, some of which have even been opened to tourists, generating revenue for local people.

The park itself, which is home to the greater one-horned rhino and is one of the few strongholds of the Bengal tiger outside India, is one of the biggest tourist draws in Nepal, with tens of thousands of visitors annually, and half the revenue tourism generates goes to communities in the buffer zone. That's been invested in public projects such as schools and alternative energy projects to reduce pressure on forest resources. The reduced pressure, along with management and conservation work, has helped boost populations of rhinos, tigers, crocodiles and elephants in and around the park.

Ivory looks better on the elephant

Elephants are one of the wildlife icons of sub-Saharan Africa, playing an important role in shaping landscapes, helping plants grow and forests regenerate, in turn providing resources for other wildlife in their footprints. They're also a big draw for tourists, which can be an important source of income for communities.

But they are facing a poaching crisis because of demand for their ivory.

Around 20,000 African elephants are killed every year by poachers. That's an average of fifty-five a day, or the equivalent of an elephant lost every twenty-six minutes.

There are now estimated to be around 415,000 elephants in Africa, but it has got to the point where more elephants are being poached than are being born, which threatens their future survival. The largest declines in numbers in the past decade have been in central Africa, Tanzania and Mozambique.

Despite an international trade ban on ivory, elephant tusks are smuggled out of Africa as part of the multibillion-dollar illegal wildlife trade, mainly to Asia, where they are carved into jewellery and ornaments which can then be bought by consumers. And some countries still have legal domestic markets for the product. This allows criminals to effectively launder illegal ivory by passing it off as legal, and stimulating further demand for more, which in turn drives further elephant poaching.

But action is being taken by a number of countries to tighten up their rules on ivory trade, with the US enacting a 'near-total' ban in July 2016. In January 2018, China, which is home to the world's largest legal and illegal ivory market, ➡

brought in a domestic ivory trade ban, a move hailed as a 'game changer' for elephants. Other parts of the world are following suit, including the UK, which is planning one of the world's toughest bans, as well as Hong Kong and Taiwan.

Other countries are being urged to shut down their domestic ivory markets, including Laos, Myanmar, Thailand and Vietnam, where elephant ivory can still be bought and sold, in order to make sure the problem doesn't just shift to other countries. Laws preventing the sale of ivory also have to be properly enforced.

There are also efforts under way to change consumer behaviour by trying to make it socially unacceptable to buy ivory. Half of Chinese people surveyed in 2017 said they had bought ivory at least once in the past, and a third had bought something in the last year.

And efforts continue to protect elephants in the wild in Africa and Asia, including training officials in forensic analysis of crime scenes and using sniffer dogs trained to detect ivory. Rangers are using smartphones to improve monitoring, landscapes are being protected from unsustainable development, and organisations are working with local communities to stop the animals coming into conflict with people.

But until people stop buying jewellery and ornaments made from elephant tusks and the markets for ivory dry up, elephants will continue to face poaching to feed the demand from consumers.

11

Pledge for the future: Make your pension an investment for the planet

Payday is a great day, whenever it falls. Hopefully it means your finances look rosier, the bills can be paid, and something, perhaps, put away for a rainy day. Lots of us don't even notice the bit that's being put away for a rainy day. If your pension is taken straight off your salary, it can feel almost as though it isn't your money, just a thing you are barely aware of as it slides off somewhere to wait until you're old.

Even if it's not a company pension, but a private one you've chosen for yourself, you probably don't think about the money often. It's out of sight, out of mind. It's the same with savings, if you're putting something away each month for yourself, your children, or your grandchildren. The money gets squirrelled away in a savings account or in some kind of investment product and forgotten about most of the time, just as if you were tucking notes under your mattress, or dropping coins into a piggy bank.

But unlike the squirrelled notes and coins, the money in your pension or savings or investment account, even the money that sits in your ordinary bank account, isn't just slumbering away in the dark. It's out there, in the global financial system, part of the ebb and flow of the world's money.

And some of that money is being spent on good things, invested in progressive companies that are trying to take their responsibility for resources seriously – whether they're developing clean energy or technology, or just walking a little lighter on the planet than others. But some of the money isn't being spent on good things.

So it's not just how you spend your money that makes a difference – choosing alternatives to plastic, buying energy-efficient appliances, or favouring train over plane. If you've got money in a pension, or in the bank, or in savings, you can take steps to make a difference with that money. And if you invest your money to help the next generations of your family, the last thing you want to do is to bolster the things that are going to destroy the life-support systems of the planet they're going to live on. You can begin by asking yourself something simple but important: 'What is my money up to when I'm not looking?'

It's not a trivial question, and it's one that is beginning to be asked by increasing numbers of people and institutions, because it's really important. For instance, huge amounts of money need to be shifted into renewables, low-carbon technology and energy efficiency to stop the worst effects of global warming. The International Energy Agency estimates that US$75 trillion will need to be invested over time in these kinds of things if we are to try and keep temperatures from rising more than 2°C above those of pre-industrial times.

If this amount of money seems huge, so too are the potential financial consequences of not making the move to an economy that focuses on low-carbon forms of electricity, transport, heating and industry. There will be costs ranging from unrepaid loans on harvests that fail because of higher temperatures, to huge insurance payouts for floods and natural disasters caused by extreme weather. It's not just the direct hits that could cost money; there will also be the financial consequences of slower growth and lower returns if companies, economies and countries struggle with the impacts of climate change and loss to key biodiversity and life support systems.

This threat can be hugely reduced by taking action to curb rising temperatures now. That action can be profitable too, with new markets in renewables, electric vehicles, low-carbon heating systems and other technologies beginning to ramp up.

Companies need to prepare for the future, particularly if countries follow through on their promises to take the steps necessary to cut emissions in line with the Paris Agreement to avoid dangerous climate change.

Some companies are already making the shift, such as Danish company DONG (Danish Oil and Natural Gas), which recently changed its name to Orsted, after Danish scientist Hans Christian Orsted, who discovered electromagnetism in 1820. The change reflects the company's shift from 'black' to 'green' power, focusing on wind farms, bioenergy plants and generating energy from waste sources.

Others, however, are not making the change, and investing in them could leave money 'stranded' in assets that can no longer be used in a low-carbon world. For example, if governments legislate to curb climate emissions, then investments in oil wells or coal mines and power plants will not generate the currently expected returns. The same will be true if demand for fossil fuels decreases significantly because we're largely running on green energy and driving electric vehicles.

All of this can seem rather remote when you're sitting in your kitchen, thinking about plastic packaging or growing your own veg. If anything, the financial crash in 2008 reinforced the idea for many people that global financial markets are a runaway train that governments and regulators can't control, let alone ordinary people. But money is power, power which the market has stripped away from ordinary savers and owners of capital. So if you have your money in an investment fund or pension, it's time to start taking back that power.

Pension managers have a duty to the people whose money they manage to ensure they are acting in their best interests, though often that care is judged by short-term returns. So one place to begin is to find out from your pension provider which fund your money is in and whether it's a sustainable fund. If not, you can shift it, but if you do that you should be thinking about which companies the fund invests in. What are those companies doing, and are they helping to build a sustainable future for everyone? What is the fund's strategy for where (and how) it invests?

There are new sustainability ratings for funds on the market, but these sometimes need to be taken with a pinch of salt. It's often better to do a bit of your own digging to look under the bonnet of the fund and see which companies are included, why, and what products and services these companies are producing. It should be easy to research the top ten companies for every fund, and it's a good place to start. Checking out what the companies are up to on issues like climate change and deforestation is interesting, and you'll learn all sorts of fascinating things along the way.

One of the ways of judging whether companies are helping to build a sustainable future is seeing if they've signed up to business initiatives on the environment, such as Science Based Targets. These show they are likely to be taking steps to reduce all the greenhouse gas emissions associated with their business, so that they are in line with the Paris Agreement goals. Of course, this is no substitute for examining which companies the fund holds.

Finding funds you like the look of can be an

interesting experience, and you can make sustainable investments that are also profitable in the long term.

Pension funds are beginning to take issues like climate change seriously, amid warnings from experts that it is not just a theoretical long-term risk, or simply an ethical decision, but could be a major financial headache in the coming years. Some investors, including pension funds, are even changing their policies to avoid investing in some areas. They're 'divesting' their money from the dirtiest fossil fuels such as coal, not only because shifting away from carbon-polluting energy sources is the right thing to do, but because they see the financial risks of continuing to invest in them.

In other cases, institutional investors who are not divesting are starting to demand that the companies in which they hold shares, such as big oil firms, report on the risks that climate change poses to their businesses. They're calling on firms to state how they're taking action to fall in line with the move towards cutting emissions.

If you've got money of your own to invest, and you've really got the bit between your teeth on engaging with big business, you could buy some shares in big oil or mining firms, for example. Then you can go along to their annual general meetings where ordinary shareholders have the opportunity to ask the executives of the businesses about the things that they think are important, such as what the company is doing on climate change or environmental protection. You might even get a free lunch.

But if you'd rather keep your money away from such companies, you can also find funds to invest in that aim to deliver social and environmental benefits alongside a healthy financial return. As with your pension, it's a good idea to check them out in some detail, but it's not necessary to sacrifice financial performance for doing the right thing – as these funds can deliver on both.

For most of us, the main way we take care of what money we have is to put it into a bank account, so think about your bank too and what your money is up to there. Some banks are transparent, such as Triodos Bank, which publishes on its website details of every organisation it lends to, so you can see exactly what your money is doing: for instance, funding renewable energy projects or care homes for the elderly.

There are other banks that have sustainability policies, such as avoiding lending to high fossil fuel industries or the arms industry. Or you could think about building societies or even credit unions, particularly for savings and loans. They are both owned by their members rather than shareholders and as a result they don't have to focus on returning a short-term profit to those shareholders at the detriment of the planet.

If you don't want to change banks, you can always challenge your bank on its lending policies. It may seem difficult, as just one person, to question multinational banks that control billions of dollars, pounds and euros. But a little bit of those billions is your share, and you have every right to ask what is happening to it. So get in touch, ask your bank how it is safeguarding your money against the risks posed by stranded assets and the potential financial losses of climate change, and find out how it is using your money to support the transition to a cleaner, fairer and more sustainable world.

That way, you can help start to tilt the power towards the people who put money into the system and who want to see that money used for the good of the planet and future generations. That's the real bottom line: it's our money and our power, and now is the time we should take it back.

From stranded assets to Science Based Targets

If the phrase 'stranded assets' conjures up visions of a suitcase of your money beached on a desert island where you can't get to it, you're not a million miles off the mark.

In the context of climate change, it refers to investments in fossil fuels that will not deliver the economic return that might once have been expected of them. For example, a coal-fired power plant that goes on burning coal and making money from the electricity it produces, as if there were no such thing as global warming, could now find it faces unsupportive government policies. That means people who invested in the coal plant and were expecting a return for their money won't see it.

In a key report in 2011, financial think tank Carbon Tracker warned that the world was sitting on a whole load of unburnable carbon. They estimated that only around 20 per cent of the proven fossil fuel reserves owned by private and public companies could be used before the world bust the budget for the amount of greenhouse gases we could emit and still keep temperatures below dangerous levels – which means investment in the rest of the reserves was effectively in a financial 'carbon bubble'. According to ➡

Carbon Tracker, there are already examples of coal mines, coal and gas power plants and hydrocarbon reserves which have become stranded by the low-carbon transition.

Investors are beginning to take note, and to demand that companies assess the risks climate change poses to their business.

Many companies do understand the risks and opportunities of climate change for their operations, their customers and the world, and are taking steps to cut their own greenhouse gas emissions. They know, for example, that rising temperatures will hit production of food from rice to cocoa, and that switching to a low-carbon economy will be a boon for electric car manufacturers and clean-energy companies.

More than 100 global corporations now have Science Based Targets in place. This initiative means the greenhouse gas emissions cuts they pledge their business will deliver are independently judged to be in line with what is needed to meet the Paris Agreement goal of keeping temperature rises well below 2°C above pre-industrial times.

As a result of signing up to the Science Based Targets initiative, food companies are working with farmers to reduce their emissions and technology manufacturers are making their products such as laptops more energy efficient.

Supermarket chains are switching to renewable energy and making their refrigeration more efficient. Firms that produce cleaning and personal care products are aiming for zero-deforestation from palm oil and are getting consumers to reduce the temperature of their wash. Some energy firms are even getting out of fossil fuels and into renewables, and new and exciting businesses and business models are coming on stream.

Taking action helps companies get ahead of the game on new innovation and government policies to cut greenhouse gas emissions. This bolsters the confidence investors have in their businesses, and improves their reputation with customers. It can also help them compete in the future when resources become more expensive. Ultimately, it makes companies more resilient.

And as companies are such power players in moves to curb global emissions, it's crucial they play their part.

Withdrawing money: what is divestment all about?

Divestment isn't a new idea – it's been circulating for decades in relation to putting pressure on governments, companies or institutions to change their policies or behaviour. Put simply, it involves taking money out of where it is invested if you don't agree with what the company or organisation is doing. It's most likely you'd then invest it in an alternative that you feel is more aligned with your views or what you want to achieve.

It has taken on a new life in the light of climate change, amid warnings that the majority of fossil fuel stocks that we already know about but haven't dug up yet must stay in the ground if we're going to avoid dangerous temperature rises.

Universities, pension funds, local and regional governments, philanthropic organisations, faith organisations and even banks and insurance companies have committed to divest from fossil fuels. Some of them have chosen only to focus on the most polluting fuels such as coal and tar sands, but others have gone the full distance and divested from all fossil fuels.

Campaigners for the movement estimate that hundreds of organisations with investments worth more than US$6 trillion have fully or ➡

partially divested from fossil fuels, along with tens of thousands of individuals controlling more than US$5 billion.

The divestment movement argues that shifting money away from the fossil fuels that create global greenhouse gas emissions is both financially astute and a moral imperative. The fossil fuel era is ending and investments in the sector will devalue as the world shifts to a cleaner economy, they argue. A failure to make the change will lead to dangerous climate change that will disrupt the global economy and disrupt all types of investment.

There is also the argument that institutions such as pension funds have a duty to the people whose money they manage to ensure it is invested in the best way possible so it will deliver returns in the future. At the same time organisations which divest may feel they have a moral duty to do so, such as churches who warn of the damage climate change is doing to God's creation and their fellow human beings.

While divestment may not directly change the flows of money – after all, if someone is selling shares in a fossil fuel company, someone else is buying them – it can have other important effects.

It sends a powerful signal to governments and fossil fuel companies who are not investing in clean alternatives that investors do not want to support businesses that pose a threat to the future of the planet. It shines a light on the activities of those businesses and has been successful in reducing the lobbying power of oil, gas and coal companies in their efforts to prevent climate action.

And investing in clean alternatives boosts industries that are trying to deliver the kind of things, from renewable power to electric cars, that will help meet the challenge of tackling climate change.

Norway's sovereign wealth fund: from oil to climate action?

It was only in the 1960s that the Norwegians discovered their waters contained something far more valuable than fish. It took four years of offshore oil prospecting, with most companies giving up, before oil was finally found two days before Christmas in 1969.

What they found was the largest offshore field ever discovered. This was soon followed by the discovery of other large fields which went into production.

Revenue from tax and direct ownership meant that Norway benefited hugely from the offshore finds of 'black gold'. In the first twenty years, money generated from oil and gas went towards developing the industry and the country as a whole.

But then the Norwegian Oil Fund, or the Government Pension Fund Global, was established by the government and in 1996 the first capital was transferred to the fund. It is owned by the people and managed by Norges Bank, the Norwegian central bank, and exists to generate wealth for current and future generations of Norwegians.

Years of investment in companies, government bonds and real estate outside Norway have seen the fund become worth more than US$1 trillion. It's the biggest sovereign wealth fund in the world, and owns more than 1 per cent of most of the world's shares.

The fund is run transparently, with anyone able to see what it has invested and where, and it operates on a series of ethical and environmental criteria. It excludes investment in companies such as tobacco firms, nuclear weapons manufacturers, mining companies that cause serious environmental damage, and those that abuse labour rights.

It also engages actively with the companies it invests in, holding thousands of meetings where it raises environmental, social and governance issues, and votes at thousands of annual general meetings of shareholders.

The fund now also calls on companies to report on the risks to their business from climate change.

In the last few years, the fund has taken steps to divest from companies that base 30 per cent or more of their activities on coal, or receive 30 per cent or more of their revenues from it. Dozens of companies are excluded from investment on the grounds that they are involved in the most polluting of fossil fuels.

Norway is still drilling and pumping out oil and gas, but the fund that was built on oil money has proposed divesting from oil and gas elsewhere in the world to reduce the risks of permanent falls in the value of fossil fuel assets. Though the motivation is mostly financial, to some observers it seems to show which way the wind is blowing.

12

Don't mess with the world: Stop littering, and lead a community clean-up

You're walking along a street near your home, carrying your reusable water bottle. The sun is shining and you're thinking about what you are going to cook for dinner. The weekend stretches out in front of you, with no plans except to relax and enjoy yourself, or perhaps spend time with the family, see some friends, do a bit of gardening. And then you see a plastic bottle lying in the gutter or in the grassy area by the side of the road. It's unsightly, making your area feel unkempt, uncared for. It can also make you feel like you're the only one bothering to try and do something about the plastic pollution problem.

Or you decide to take a trip to the beach at the weekend, and there's a wide expanse of silver-white sand under a blue sky with some wisps of cloud on the horizon, and the air smells of salt and seaweed, and you dig your bare feet into the sand. But then you look along the shoreline, you can see that there's rubbish and debris everywhere. And all the wonderful feelings that rushed in as you stepped down from the road or cliff path and absorbed the scent and the sound of the beach are shot to pieces by the sight of all that human rubbish. You can't help but remember what you've read about what's happening to our oceans, and rather than feeling humbled by the vastness of the ocean you feel anger about the state it's in.

Now you might pick up a stray plastic bottle and carry it along the street or beach in search of somewhere to dispose of it properly. Or you might not because frankly you don't know where it's been and it's a bit dirty or smelly. Even if you do pick up a piece of litter, you're only going to spot another bottle or can, a cigarette butt or disintegrating plastic bag a few metres further on. And another, and another. It's a task that's too big for one person to take on.

When it comes to caring about the environment, it can be all too easy to feel that even though you're doing your bit, your actions don't matter and it's all pretty hopeless. It's enough to make you give up altogether. Or you can end up thinking that you've

done your part, so why should you do anything else, anything that helps anyone else?

But what if there was a group of you, all wrapped up against the elements or wearing hats in the sun, armed with gloves and bags to put the rubbish in? You could work your way along the beach, or through a patch of green space or a bit of countryside, picking up all that unsightly litter and chatting as you go, really having a proper conversation with someone you didn't know this time yesterday. How often does that happen?

Or you could bring along your friends or family to join you, collecting all the rubbish you find while enjoying the views, the fresh air and the sense that you're not alone in fighting this fight. You could say you're the biggest team in the world.

What you collect on a single beach can form part of a nationwide survey of rubbish found along the coastline, to give an idea of whether the problem is getting worse or not, and what kinds of items – from cotton buds to wet wipes and plastic forks – are washing up on the nation's beaches. That information, gathered by groups of normal people spending their weekend picking up litter in groups, can then be used to put pressure on governments and industry to take more action on issues like single-use plastic.

This kind of information gathering has become known as 'citizen science' and it's a really powerful way to make your contribution count towards a greater understanding of the world around us, and to prompt action by governments and corporations, who are forced to respond to citizens and organisations armed with knowledge. It is knowledge a few members of staff at a charity, environmental organisation or academic institute would take years to gather, years we don't have if we're to change things before it's too late.

Take water, for example. You can take action to cut your water use and engage with your water provider about what they're doing to protect the environment where they source the water from. But you might also be able to take part in a 'bioblitz', in which scientists team up with volunteers to comb through a specific area or habitat looking for all the species that can be found in a short period of time – riverside plants, birds and creepy crawlies. You're recording, in effect, the whole diversity of life in a small patch of the Earth at a moment in time. Bioblitzes can provide scientists with a snapshot of whether life is thriving in a given place. And if you've carried one out by a river or stream or wetland, that information could be used to demand action to improve habitats that are struggling because of the impact of invasive species, or because too much water is being removed. Or it could help identify areas where the rich nature needs greater protection.

There are a multitude of citizen-science projects out there, from butterfly recording to photographing mountain-top haze. Often you can help out with information that you see on a walk in the countryside or even what's going on in your back garden. In Australia people are encouraged to record sightings of the elusive platypus. Meanwhile, in the UK people can contribute their sightings of the changing seasons, from the bud burst of oaks to the first appearance of butterflies. These observations feed into a set of records that dates back to the early eighteenth century. The idea now is to build up a picture of how the seasons are shifting due to warming temperatures, and the information is being used in scientific research about the impact of climate change on wildlife. In India a similar scheme has been set up that encourages people, including children, to choose a tree and visit it weekly to record when it flowers and sets seed, to see how that is changing. Then there are schemes that ask you to submit photographs of things, to monitor anything from what puffins are feeding their chicks on the Scottish coast to air pollution in the Appalachian mountains.

You can even help out from your own home using your laptop, by analysing huge amounts of data recorded by drones or other automatic systems. This could be counting and reporting the number of penguins in a photo from Antarctica or identifying bits of plastic waste from shots filmed along the shoreline. Some citizen science lets you help simply via the processing power of your laptop when you're not using it.

But most citizen-science projects take us closer to nature, giving us the opportunity to contribute to something far larger than ourselves. And these simple acts can make a big difference.

They send a message to those in power that people care, and that they have gathered knowledge so they can change things for the better.

We send messages through how we behave or what we buy: if lots of people carry water bottles, for example, the message gets through that we need water fountains in public spaces. If more people choose clothes made from sustainable cotton, companies will seek out better materials for their clothing brands.

But our messages can be more direct. We've seen how you can effect change by getting in touch directly with companies. That's just as true when it comes to your elected officials.

Whether it's your local mayor or the prime minister, they represent you, and they want your vote. So if you want better local cycling facilities, ask your local councillor to meet and discuss it with you. If you want stronger policies or laws to push for greater amounts of clean energy, more quickly, get in touch with the person who represents you on a national level, such as your MP. Write to them, go to public events or hustings when people are standing for election and ask a question, or arrange a meeting, to ask them about the things you care about. These are the men and women who will shape the nation your children will grow up in, and they will enter environmental agreements with other leaders from overseas: don't leave it up to others to control the conversation. To be a responsible global citizen, it's important that you hold them to account in the polling booth too, so make sure you vote.

You can also join a campaign for a one-off issue, like action to protect a species or habitat, or a campaign group working on long-running issues such as climate change. Go on a march, sign a petition, send a postcard, join a social media campaign or respond to a consultation. If you're with others on a march, you'll see and hear how many other people there are like you, wanting change, determined to achieve it. If you sign a petition or get involved in a campaign you'll be able to track its progress and what it achieves. Sometimes the results can be stunningly fast, and when they are, use that to get enthused for the greater, harder, more intractable challenges. In all these ways, directly or indirectly, you are creating the space that governments, councils and companies need to be able to take action, giving them the mandate to deliver change.

More than that, you are helping drive a revolution, one that is needed to secure a sustainable future for the planet and everything living on it. That's not such a small thing, is it?

The world's biggest beach clean

People around the world have joined community beach cleans in their local area to try and rid them of rubbish, but perhaps no beach clean has been bigger than the effort to clean up Versova Beach in Mumbai of more than 4,000 tonnes of plastic, glass and rubbish that had accumulated there.

Indian lawyer Afroz Shah was awarded the UN's top environmental prize, the Champions of the Earth award, for his work to inspire hundreds of volunteers to join him in cleaning up the plastic bags, cement sacks, glass bottles, clothing and shoes which entirely covered the 2.5km beach.

Shah and an elderly neighbour began the crusade to clean the beach where the rubbish lay shin-deep in places, picking it up themselves, by hand. Shah encouraged other people to take part by going round the local area and explaining the damage caused by marine litter.

The beach clean was turned into a 1,500-strong community action each weekend, with everyone from schoolchildren to Bollywood stars lending a hand until the stretch of sand was cleared.

Shah has said humans need to reignite their bond with the ocean, and that he wants to inspire people in coastal communities around the world to take action against marine pollution.

On Versova the awe-inspiring effort has really paid off: by 2018 the beach had been cleared enough to enable eighty olive ridley turtle hatchlings to make their way towards the sea, guarded by volunteers who made sure they got into the surf safely.

Others are also tackling the problem on their own shores, such as on the Indonesian island of Bali, where officials deploy trucks and cleaners to remove the flood of marine litter from beaches. In addition to the government response, in 2018 a coalition of local people, organisations, businesses and hotels under the banner 'Bali One Island One Voice' organised a second annual beach clean. They called on government, locals, schools and tourists to take part and managed to spur 20,000 people to join 120 clean-up operations around Bali. Between them they picked up 60 tonnes of rubbish – including straws, plastic bottles, plastic bags, shoes, glass, cigarette ends and fishing gear.

And then there's 'plogging', a workout trend which sees groups of people picking up litter while they're out for a run. It kicked off in Sweden, but has now gone global, in part thanks to social media.

Of course, picking up litter from beaches or the countryside is not enough on its own, given the scale of problems like marine plastic pollution, even if it is done by teams or hundreds of people rather than just individuals. But these actions are vital as they raise awareness of the situation and galvanise action to address the root causes, as well as just cleaning up some of the mess.

How to blossom at school

Getting out and doing some gardening helps children to think about the whole system of the natural world around them and how it is all connected and interacts. For example, how if we don't have bees and other pollinators, it will have impacts on important things like food production.* It can also help them learn about nature in a hands-on way, and see how planting for wildlife can boost our 'minibeasts', butterflies and birds.

Growing their own food can also inspire young people to think about where our food comes from and what can be done to produce and consume

* WWF's Plant2Plate campaign focuses on what can be done to produce and consume food in a sustainable way that is not harmful to our planet. There are some fantastic curriculum linked resources and activities for Key Stage 1 and 2 pupils, which can be downloaded at: www.wwf.org.uk/get-involved/schools/school-campaigns/plant2plate

food in a sustainable way. There are practical ideas out there for the best seasonal fruit and vegetable for schools to grow so they can get good harvests in before the summer holidays. And once they've grown the food, the children can head to the kitchen, put on their chef's hats, and start coming up with dishes that use those ingredients. This can also boost health, which is important at a time when increasing numbers of children are becoming obese or overweight. Being in the garden is healthy too, providing fresh air and physical activity.

School gardening plays into lots of different parts of the curriculum, such as design and technology, science, maths or the arts, as children map out the flower and vegetable beds, study how they grow and take inspiration for creative-writing classes.

But it's not just about what they learn in formal terms, it's about what growing things can teach children in the widest sense, including young people who don't necessarily thrive in the more traditional setting of the classroom. Research and evidence from schools show that growing things gives children a sense of responsibility, and helps them learn teamwork and communication as they pull together to transform a patch of earth into a thriving vegetable plot or wildlife garden.

School attainment can go up, as can levels of attendance, self-esteem and confidence. The children blossom along with the garden.

And it's not just for the pupils. Schools frequently call for help from parents, grandparents, volunteers and gardening clubs to help the gardens grow, while local businesses can offer their support and children are often encouraged to sell their produce to learn financial skills and so people can eat fresh, local food. A school garden or allotment can strengthen communities across generations and ethnic backgrounds. They begin by planting seeds in the ground, and end up growing a community.

The People's Climate March

In September 2014, world leaders met in New York for a summit organised by Ban Ki-moon, who was United Nations secretary general at the time, to push towards securing a global deal on climate change in Paris the next year. It was the first such meeting of leaders on climate change since the failed climate talks in Copenhagen five years earlier.

And the people sent the leaders a message: this time they wanted action.

Spurred on by activists, from the streets of Adelaide, Australia, to playing fields in ➡

Gloucestershire, England, people came together to call for climate action from their leaders.

More than 2,000 events took place in 150 countries around the world as part of the People's Climate March, with dancers and campaigners in fancy dress in the sunshine in Bogotá, Colombia, and marchers braving driving rain in Rio de Janeiro, Brazil.

A 'Climate Action Now' banner was hung on Australia's Blue Mountains, and people in Delhi marched with mocked-up wind turbines and solar panels strapped on to themselves. Thousands stood to spell out 'Beyond Coal and Gas' to the sky in Sydney Harbour, and took to the streets of Paris and the beaches of Spain.

In London, up to 40,000 people turned out, carrying banners that said 'Renewables Rock' and 'For the Love of Polar Bears and Rhinos' and chanting, 'What do we want? Clean Energy. When do we want it? Now.'

Puppet giraffes, birds and zebras were joined by people dressed as pandas, and parents with their children in pushchairs. A little girl dressed as Elsa from *Frozen* held a banner that said 'Keep Things Frozen'.

The biggest march of all was in New York itself, the centre of the action, where organisers estimated 310,000 people lined the streets between Manhattan's skyscrapers, many of them coming from across the US to take part. There ➡

was a moment of silence before an outbreak of cheering to sound the alarm for the climate. A Noah's ark bobbed through the sea of people and sunflowers, carbon bubbles floated past, scientists carried technical signs spelling out rising temperatures, brass bands played and people wore their indigenous dress. Even Ban Ki-moon was there, joining demonstrators as they sent the message to world leaders that urgent action on climate change was needed.

And they kept marching for the next year, right up to the Paris climate talks in December 2015. Even when they couldn't march, because of security concerns, they left thousands of pairs of shoes in Paris's Place de la République, including a pair from Pope Francis, to show they would have marched, they still wanted action.

Countries had laid the groundwork before the Paris talks, each setting out their national plans to counter the climate threat. They were helped by changes in technology and reductions in costs that make the switch to clean tech more likely, together with clearer messages than ever from the planet that the climate was changing, and the political will to do something about it – spurred on by public pressure.

A deal was agreed, and people power played no small part in it. Now to make it a reality, we have to keep marching.